FLORIDA STATE
UNIVERSITY LIBRARIES

NOV 25 1997

TALLAHASSEE, FLORIDA

A CLAIM TO LAND
BY THE RIVER

A Claim to Land by the River

A Household in Senegal
1720–1994

Adrian Adams
and *Jaabe So*

OXFORD UNIVERSITY PRESS
1996

Oxford University Press, Walton Street, Oxford OX2 6DP
Oxford New York
Athens Auckland Bangkok Bogota Bombay
Buenos Aires Calcutta Cape Town Dar es Salaam
Delhi Florence Hong Kong Istanbul Karachi
Kuala Lumpur Madras Madrid Melbourne
Mexico City Nairobi Paris Singapore
Taipei Tokyo Toronto
and associated companies in
Berlin Ibadan

Oxford is a trade mark of Oxford University Press

Published in the United States
by Oxford University Press Inc., New York

© Adrian Adams & Jaabe So, 1996

All rights reserved. No part of this publication may be reproduced,
stored in a retrieval system, or transmitted, in any form or by any means,
without the prior permission in writing of Oxford University Press.
Within the UK, exceptions are allowed in respect of any fair dealing for the
purpose of research or private study, or criticism or review, as permitted
under the Copyright, Designs and Patents Act, 1988, or in the case of
reprographic reproduction in accordance with the terms of the licences
issued by the Copyright Licensing Agency. Enquiries concerning
reproduction outside these terms and in other countries should be
sent to the Rights Department, Oxford University Press,
at the address above

British Library Cataloguing in Publication Data
Data available

Library of Congress Cataloging in Publication Data
Adams, Adrian, 1945–
A claim to land by the river : A Household in Senegal,
1720–1994 / Adrian Adams, Jaabe So.
p. cm.
Includes index.
1. Ethnology—Senegal.   2. Soninke (African people)—History.
3. Soninke (African people)—Agriculture.   4. Gajaaga (Kingdom)—
History.   5. Agriculture—Social aspects—Senegal.   6. Agricultural
development projects—Senegal.   7. Senegal—Social conditions.
8. Senegal—Social life and customs.   I. So, Jaabe.   II. Title.
GN655.S3A325   1996
306'.09663—dc20   96-5771

ISBN 0–19–820191–5

1 3 5 7 9 10 8 6 4 2

Typeset by Best-set Typesetter Ltd., Hong Kong
Printed in Great Britain
on acid-free paper by
Biddles Ltd., Guildford & King's Lynn

*The stone which the builders refused
is become the head stone of the corner.*
                                    Psalm 118

Prologue

Most stories about Africa today are told by bystanders, and chronicle disaster under different names. This book tells instead one of the African stories interrupted; for its own sake, and also in the hope of shedding new light upon the others. The story spans three centuries, but is yet unfinished. It is a story for the future.

*

The words of Adrian Adams were translated into Soninke, then read out. The words of Jaabe So and all others were recorded and transcribed in Soninke, then translated.

*

There is a game some children play, called 'Scissors, Paper, Rock'. Each brandishes his fist in the air, once, twice, and at the third count brings his hand down in one of three possible signs: Rock, the clenched fist; Paper, the hand held rigid, fingers joined; Scissors, the index and middle fingers spread. Paper wraps Rock, Scissors cut Paper, Rock breaks Scissors; those are the rules.

The winner of a round taps the loser on the wrist, and the game continues; a game to pass the time, a game no one wins in the end. Let it serve as an emblem of a deadlier game, in which we are trapped between the rock of memory, the paper of lies, and the scissors of oblivion.

Contents

List of Maps	x
Maps	xi–xiv
Part I: Rock	1
1. By the River (*c.*1720–1933)	3
2. Inland (*c.*1720–1938)	51
3. Abroad and Home (1937–1968)	72
Part II: Paper	105
4. Paris (1968–1973)	107
5. Hope Traduced (1973–1975)	114
6. Behind the Lines (1975–1978)	132
7. Epitaphs (1978–1981)	157
8. A New Beginning? (1981–1984)	177
Part III: Scissors	199
9. Mock Reforms (1984–1986)	201
10. What Private Sector? (1986–1989)	228
11. Real Cuts (1989–1991)	250
12. The End (1991–1994)	262
Abbreviations	290
Glossary	291
Index	293

List of Maps

1. Towns of Gajaaga — xi
2. Riverside land, upstream from Kuŋani — xii
3. Riverside land, downstream from Kuŋani — xiii
4. Kuŋani farmlands, riverside and *jeeri* — xiv

Maps by Charles C. Adams.

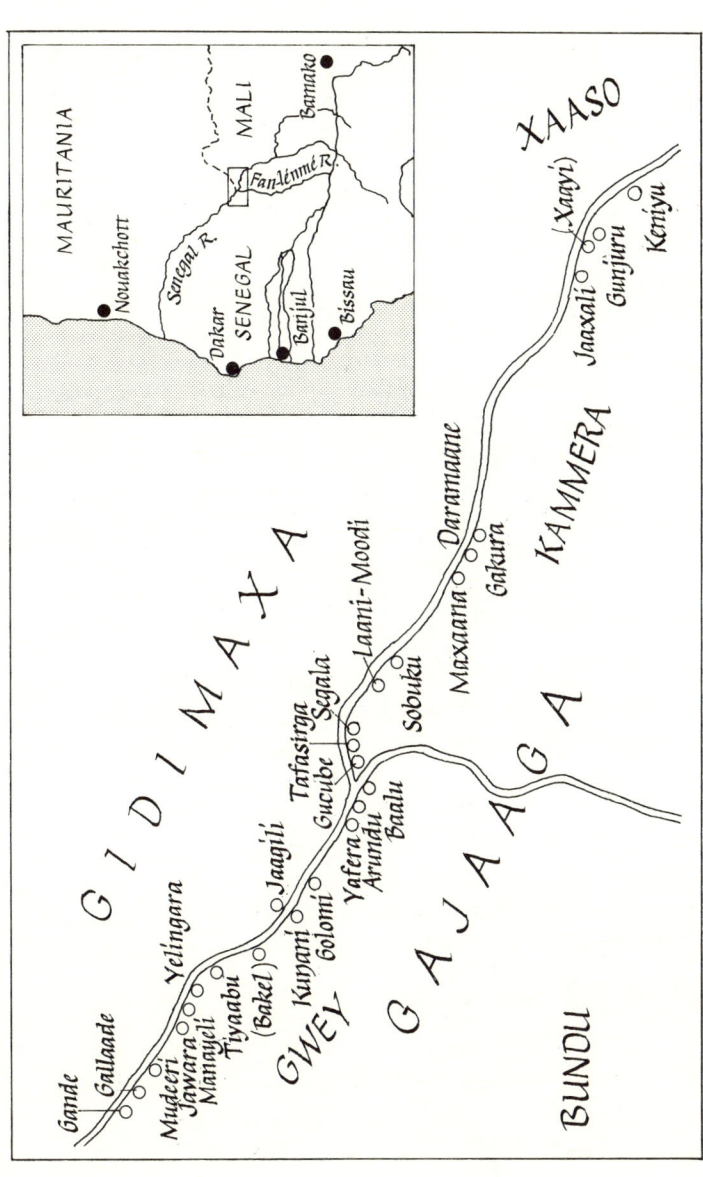

Map 1. Towns of Gajaaga

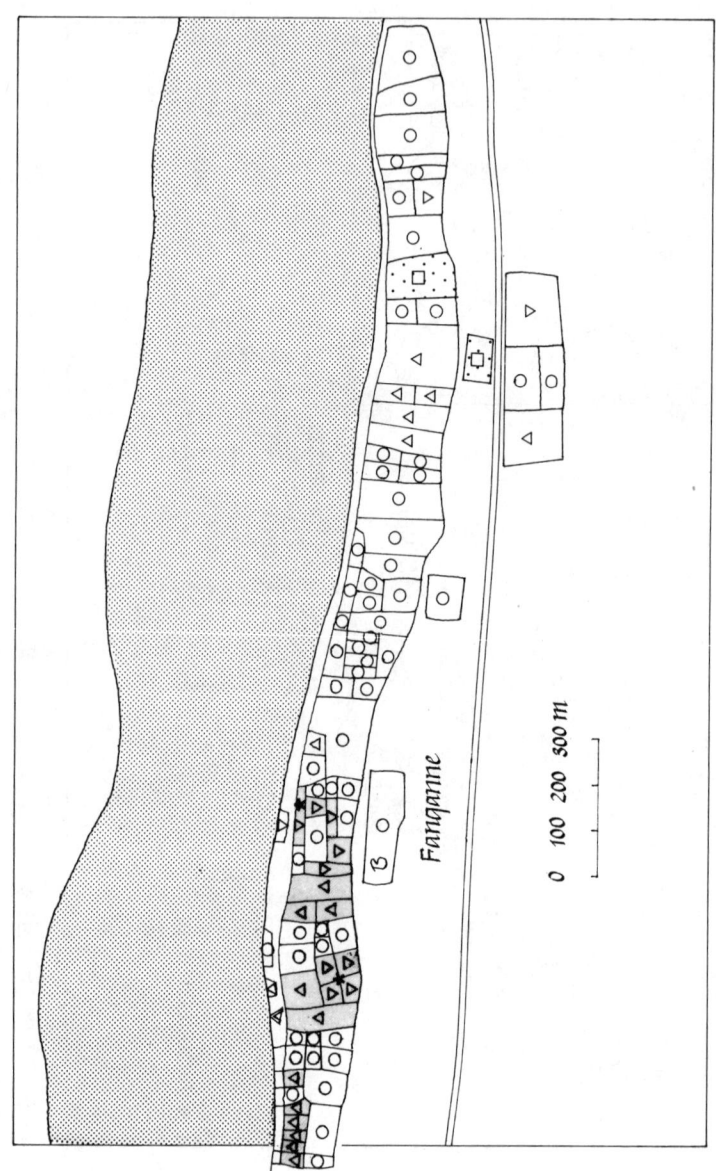

Map 2. Riverside land, upstream from Kuŋani

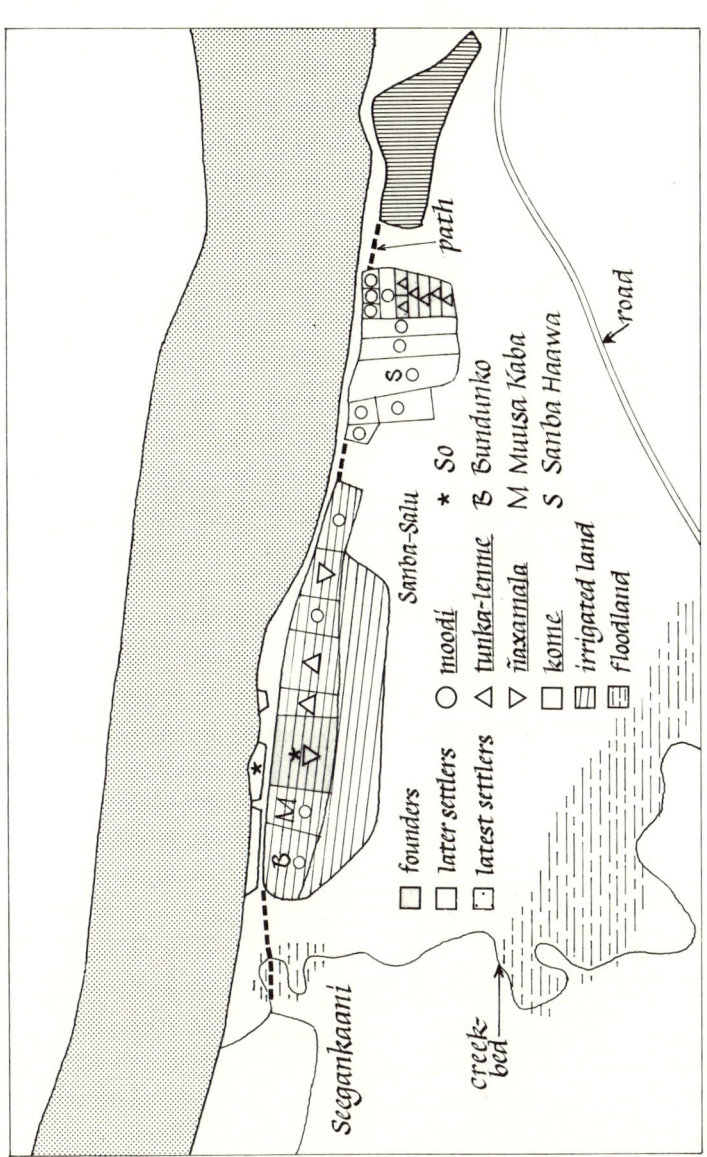

Map 3. Riverside land, downstream from Kunjani

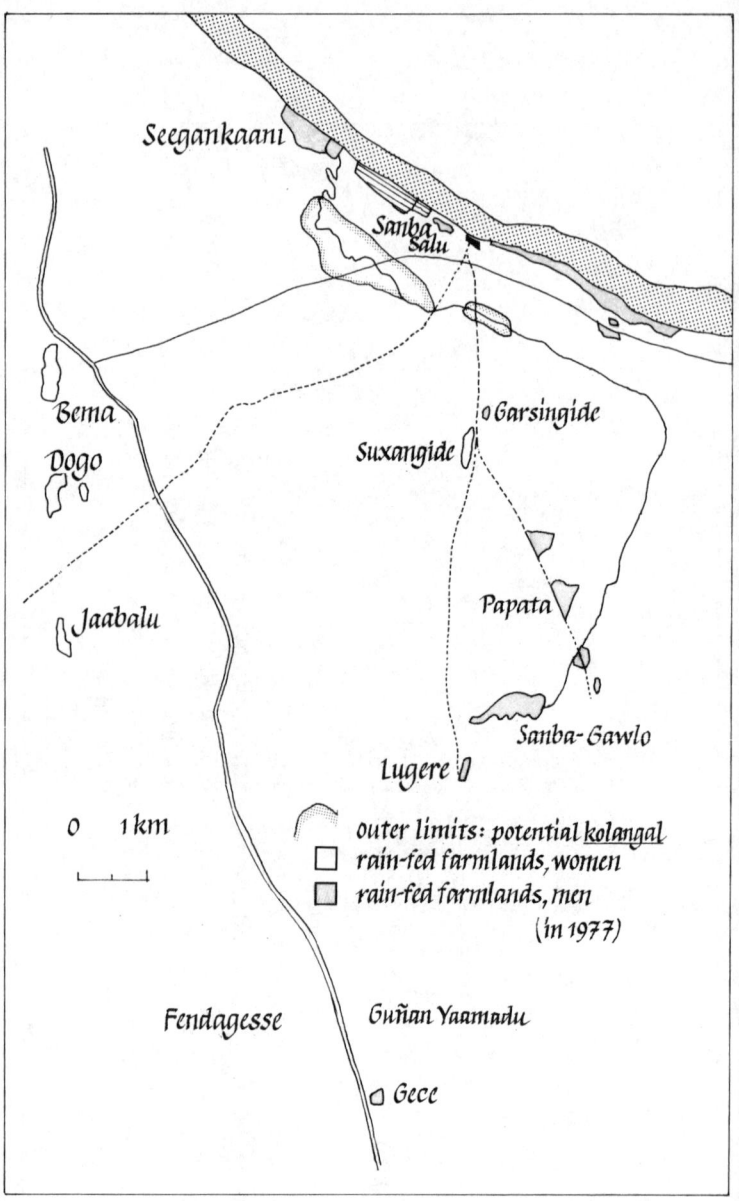

Map 4. Kuŋani farmlands, riverside and *jeeri*

PART I

Rock

1

By the River
(*c*.1720–1933)

Buleeli Denba

Adrian Adams

Begin with these fifteen acres: bounded by the river, an outcrop of rock, and the deep bed worn by a rain-fed stream flowing to meet the river as it rises.

The bed of the stream is dry. At the edge of the land, a steep bank drops to the river. The river is low; from here to the north bank, it lies like a clouded mirror. It comes from the east, but you cannot see it flow. Inland, further south, a range of low hills leads to the near horizon.

The bed of the stream is in darkness. A mile upriver, light strikes twin spires rising from a cluster of trees. Beneath them lies a little town; people are returning home from prayer, along narrow lanes between earthen walls. A few miles downriver, the light strikes a white fort on a hill.

Last night, the year's first rain felted the hard earth, releasing a strong sweet scent of life renewed into the milk-warm air.

The land lies bare between bushes and tree-stumps; it has not been farmed for years. When it was last cleared, bullets were found buried there: bullets once fired from the far side of the gorge at troops sent from the fort on the hill.

A woman is buried somewhere on the far side of the outcrop of rock. When she came here from the east with her husband, there was no fort, nor any town nearby.

Eight generations lie between that woman and the man who now speaks.

Jaabe So

O sere fana, our ancestor Buleeli Denba So lived in Maasina. He had two sons who went off to the the war between Maasina and Gurdaame.

These were two kinds of Fula, who were always fighting each other; the finest horsemen in Africa, finer than the Bacili.

Haruna returned and asked after Lasana. They told him that Lasana was seen on the battlefield, alone against three horsemen. Haruna went to Gurdaame; he killed eight people, men and women, and he himself was killed.

Buleeli Denba said: 'Ever since my grandfather came to Maasina, the men of my house have been dying in battle. Now my house is ended; only my daughter Juumu is left.'

He gathered his cattle and horses, and left. He came to Baalu, Old Baalu in Gidimaxa, and halted there with his herds, his horses, his daughter and his wife Jeela, the mother of the two young men. Along the way he had taken another wife, Seega, from Xaayi-Xaaso; in Baalu she became pregnant and gave birth to a son, Suraharta.

The people of Baalu wondered what sort of a man this was. They saw that he had a large herd of cattle, that he felled large trees and hollowed them out to make drinking-troughs. When they saw him do this, they said he must be a *sakke*, a woodworker. So they named him *Baalu-Sakke*.

In those days, the Bacili horsemen rode out every morning to collect tribute: grain, cowrie shells, cattle and sheep, and the blocks of salt brought by traders from the country of the Moors. One day Buleeli Denba went off into the bush, leaving his daughter Juumu alone by their well. The Bacili came past and saw her, alone with her dogs. They said they wanted to water their horses. Juumu said no. One of them took hold of the rope and began to draw water, while another held Juumu.

Juumu became angry. She ran and seized her father's gun. It was a powder-loading gun, with one bullet in it. She fired; the bullet struck a young man named Siliman, the son of Tunka Sanba's sister. As he fell, he seized his gun and fired at Juumu. The bullet struck her in the thigh. The horsemen said: 'You shouldn't have shot her, she's a woman.'

They went to the bush to fetch some leaves of a tree called *tefe*. When they came back, they chewed the leaves and applied them to the wounds of Juumu and the young man. They took Juumu back to their camp to remove the bullet; but it had passed clean through her thigh and come out the other side. When they arrived, the young man was dead, but Juumu was alive.

By the River (c.1720–1933)

When her father came back, he found blood, and no Juumu. He rode off to the Bacili camp to ask for her. The Bacili said: 'Here she is. She's killed a man.' Tunka Sanba said: 'We'll take care of her here.' Buleeli Denba said no. He left with his daughter, and tended her until she recovered.

One day, Tunka Sanba went to see Buleeli Denba. He found him lying in his rope hammock, surrounded by his dogs, with a gun across his knees. In those days, his herds were guarded by trained bulls; not even lions would attack them.

He said: 'Bacili, there are so many of you; have you come all this way just to see me?'

Tunka Sanba told him that he must pay tribute. Buleeli Denba said that he had nothing to be taxed; no gold, no cowrie shells. Tunka Sanba said: 'You have many cattle.' Buleeli Denba replied: 'Ask the bulls if they'll agree to go.' Tunka Sanba laughed. When he returned home, he said that was a man who should be with them.

They said: 'Every year we levy tribute, in gold, cowries, money and salt, and divide it among all the Bacili: Suntunkaara, those of Laani; Maxankaara, those of Maxaana; Jankamenkaara, those of Tiyaabu. But when everything has been shared out, they say it's not been done fairly, and they kill the person responsible. We need someone from far away, like *Baalu-Sakke*. He was a horseman; they won't kill him. Let us ask him to do the sharing-out each year.'

They asked Ñaŋaane of Baalu. He said: 'No, every year you kill whoever does the sharing.' But Buleeli Denba said: 'They won't kill me, nor my horse.' So he agreed, and set out. He divided everything up. When it was all finished, he said to Tunka Sanba: 'There you are; may I leave now?'

The Tunka said: 'Stay a little longer.' He said: 'No, I've my herd to see to. Every year, when you share out tribute, there's someone killed. Let everyone know I'll be leaving this evening, with my horse, my dog, and my gun. I'm not running away. If anyone is against the way I've done things, let him come.'

He left that evening. That was when the Bacili first heard the name of *Baalu-Sakke*.

The second year they came again to fetch him; and also the third. The Tunka decided to ask Ñaŋaane to let them have him. It was then that a soothsayer told Buleeli Denba: 'If you stay in Baalu, your family will not be a large one.' He went to Maxaana; the Bacili gave him the name of *Manga-Sakke*.

While Buleeli Denba was in Maxaana, Muulu Geyi came to fish on the river. Tunka Sanba told him to pay tribute. Since he had nothing to pay with, Buleeli Denba suggested he be allowed to continue fishing, and give them fish every day.

Buleeli Denba and Muulu Geyi became friends, and together they asked the Tunka for a place of their own to settle in. For Buleeli Denba foresaw that one day there would be strife among the Bacili.

At the time when they were given land, Gajaaga was ruled from Maxaana. When our town was founded, the only other town near here was Gucube. After our town was founded, the Bacili factions—Suntunkaara, Maxankaara, Jankamenkaara—separated, and those of Tiyaabu came here. Afterwards, they settled at Kaajo, a place between where Manayeli and Tiyaabu are now. When Bakel was founded, Tiyaabu was at Kaajo: both Tiyaabu and Manayeli together, because the men of Manayeli were Tiyaabu's warriors. The men of Tiyaabu and their *mangu* went together into battle. At that time there was no Jawara, no Mudeeri, no Gallaade, no Gande. Tiyaabu, Kuɲani, Gucube: that was all there was by the river, with Guñan and Bema inland.

There was an evil spirit in the place where Buleeli Denba and Muulu Geyi first settled; they moved across the river, then returned to this side a bit further upstream. When two passing Muslim clerics named Sawo and Gaku helped them exorcize the evil spirit, they returned to their first choice, a rocky hill above the river. The settlement's first name was Gurel Hayire, 'the town of rocks' in Pulaar. Buleeli Denba said that Muulu Geyi should be the town head, the *debegume*; he stayed by the river to fish, while Buleeli Denba often left to follow his cattle.

Buleeli's wife Jeela died while they were living across the river. When he returned to this side, he had only one wife: Seega Sakiliba, a Xasonke who loved cattle. She was a *tege* woman. The only *tego* ever to settle in this town were women married into our family. The town's *jinna* was a blacksmith, and didn't like *tego* men.

They had a large herd; when they settled here, the cattle could not be kept nearby because of flooding during the rains. So they asked the Bacili for another place on higher ground. The Bacili gave them land further downstream, light soil lying between outcrops of rock, Buleeli

> Tiyaabu fana, *the first Tiyaabu was between Gucube and Tafsirga*
> *The second Tiyaabu was where the ruins* (kaajo) *are, between the*
> *present Tiyaabu and Bakel*
> *Then they went to Sanba Uri's* yerro *tree*
> *Sanba Uri crossed the river and founded the Tiyaabu of today*
>
>> (From a Bacili praise-song recited by Sanba Jaare Daraame of Gande, recorded and transcribed by Abdoulaye Bathily, Bulletin de l'Institut Fondamental d'Afrique Noire, *31, 1969*)

Denba stayed in Gurel Hayire, and Seega lived out in the bush with their slaves. Sometimes they would fill whole troughs with milk. In those days, milk had little value.

People began to call the place where she lived, *Seegan kaani*, 'Seega's dwelling-place'. They would say: 'Are you looking for milk? Go to Seega's place. Are you looking for a feast-day ox? Go to Seega's place.' That was how the name Seegankaani began. Before she died, Seega asked to be buried there.

The bush was dangerous then, with many wild animals. When she died, they buried her at Seegankaani. Her husband kept watch; they lit a fire over her grave, to keep the hyenas away, and nothing harmed her. That was a very long time ago.

Adrian Adams

Where? Seegankaani lies between the river, an outcrop of rock inland, and the deep bed worn by a rain-fed stream flowing to meet the river as it rises. Gurel Hayire is a mile upriver; there are houses there now, trees in whose shade people gather, and the twin spires of a new mosque.

Tunka Sanba gave Buleeli Denba and Muulu Geyi two and a half miles of land along the river: half a mile downriver from Gurel Hayire, and two miles upriver. It did not include Seegankaani, nor the land just upstream from Seegankaani on the other side of the rain-fed stream. It was his to give, as ruler of the kingdom of Gajaaga: seventy-odd miles of the left bank of the *Fan-xoore* or Great River.

Gajaaga was divided into two provinces. Three of the four royal towns which each in turn named their senior elder Tunka, Laani-Tunka, Maxaana, and Keniyu, were in the more populous eastern province of Kammera, upstream from the confluent with the *Fan-lenme* or Little River; with several towns of Muslim clerics, Gakura,

Laani-Moodi, Daramaane, Jaaxali and Gunjuru. Seegankaani and Gurel Hayire were in the less populous western province of Gweyi, downstream from the *Fan-lenme*. The one royal town of western Gajaaga, Tiyaabu, seems to have been founded during Buleeli Denba's lifetime by a Bacili clan from the east.

Buleeli Denba never farmed at Seegankaani; no one from Gurel Hayire was to farm there for two hundred and fifty years. It lay outside the boundaries of the land given him and Muulu Geyi. The first of their land to be brought under permanent cultivation lies just upstream from the town. They kept the land nearest the town and the river, and granted the land just beyond it to new arrivals: people from Jaaxali.

This land is called Fanqanne, which means 'the river's edge', not in Pulaar but in Soninke. It is a long narrow rise of land; bone-hard now at the dry season's end, marked only by dry flattened maize-stalks and branches of thorn-bush scattered by the wind. But every inch of it is owned, and the thorn fencing is renewed each year along known and guarded boundaries.

All the Fanqanne land now owned by the descendants of Buleeli Denba and Muulu Geyi lies in the first two-thirds of a mile along the river. In the next third of a mile, several adjacent plots belong to households which four or five generations ago were one, households which in the town are also grouped together, forming most of the quarter named Tanjankunda, and that part of the Silimana quarter that lies next to Geyga. This suggests that these were among the first of the major group of settlers to ask for land: Muslim clerics from eastern Gajaaga, most of them named Tanjigoora or Jaaxo, two versions of the same name. Through them the town soon acquired its Soninke name, Kuŋani. Some said, though, that it would be more auspicious to call it *giden debe*, 'the town of rocks'.

When? There are eight generations between Buleeli Denba and today. If one allots thirty years to each generation, and longer in the two known instances where the surviving son was begotten in old age, Buleeli Denba may have been born in about 1680, and Gurel Hayire founded in about 1720.

Al-Haji Sanba Jimmera

Tanjigooran su, the Tanjigoora are all descended from Ali Buna ba Taalibi. Ali Buna ba Taalibi married a woman who bore Mohammadu Hanafi. Mohammadu Hanafi left Mecca; he crossed the Nile, and journeyed until he arrived in Soxolo. In Soxolo he met a woman named

> The city of Ghana consists of two towns in a plain. One of these towns is inhabited by Muslims. It is large with a dozen mosques in one of which they assemble for the Friday prayer... Around the town are wells of sweet water from which they drink and near which they cultivate vegetables.
>
> The royal town, called al-Ghaba ['the grove'] is six miles away [from the Muslim town]. Between these two towns are contiguous habitations... The king has a palace and conical huts, surrounded by a wall-like enclosure. In the king's town, not far from the court, is a mosque where Muslims pray when they call upon him... The religion [of the people of Ghana] is paganism and the worship of idols.
>
> *(Al-Bakri, 1067/8, Description of North-West Africa, quoted in Nehemia Levtzion, Ancient Ghana and Mali, 1973)*

Moni. She had a child just weaned; her husband was dead. Mohammadu Hanafi married Moni.

Moni's first child was named Manga. She bore Mohammadu Hanafi a child named Satanga. Since their mother's name was Moni, they were called Satanga Moni and Manga Moni. The Bacili are descended from Manga; the Tanjigoora are descended from Satanga.

Mohammadu Hanafi brought up Manga and Satanga. Manga refused to adopt the Tanjigoora religion; he chose to follow his father's fetish cult. Satanga followed his father's path and studied Islam. Manga the *tunkan-yugo* or fetishist king, and Satanga the *moodi* or Muslim cleric, both grew up and had children. They settled in two towns not far from one another. Manga was in his own country, but the Tanjigoora had their own town. The Tanjigoora often went to visit Manga, and Manga's people also visited them.

One day when a Bacili had just died, a Daraame was present. He said: 'What! Are you Tanjigoora going to bury him without washing him or saying prayers?' They said: 'They're nothing but unbelievers.' Daraame Kanji took the body, washed it, and said prayers. While he was praying, the Bacili said: 'You Tanjigoora, we are sons of the same mother. Alive, we are yours; but when we die, you give us to the Daraame Kanji.'

Everyone departed from the east. The Tanjigoora went to see a learned man of God, Al-Haji Maxoore Silla. He welcomed them and asked them what they wanted; they said they were looking for a place to settle, and he agreed to help them. He wrote a charm for them, and said: 'Tie this to a walking-stick made of a branch of the tree called *banɲe*. The stick will guide you; settle wherever it thrusts itself into the ground.' They travelled with the stick until they reached Old Jaaxali, near Xaayi. That's where they settled.

They multiplied there. Their ancestor MaJaaxo begat Madi Xulle. Madi Xulle's name was Mohammadu; his mother's name was Xulle Tanja, and that is why his descendants are called Tanjankunda. MaJaaxo also begat Sileyman, whose descendants are called Silimana, and Lasana, whose descendants are called Lahasenikunda. Those two were sons of the same mother.

The difference between them and the people of Xoje lies in the story of Marian Silla. Marian Silla had never borne a child, and she had stopped seeing what women see. Saying that she had grown old, she went to ask her mother's brother for his daughter Marian Daraame; she would marry her to her husband, that she might bear him a child. Her uncle gave her his daughter. But one day, while she was praying, what women see suddenly reappeared. A woman relative of hers told her: 'Your prayer is spoilt.' She said: 'Are you mad?' The other said: 'Just look.' It was said that if she were to bear a child, it would be named Sallifo, 'thing of prayer'. She bore a son: Sallifo Sanba, from whom are descended the people of Xoje.

Al-Haji Jaaje Jaaxo

Bacilini joɲa Soxolo, the Bacili came from Soxolo. We Tanjigoora were also in Soxolo; we left Soxolo to settle in Gunjuru and Jaaxali. Even today, Jaaxali is near Xaayi; Gunjuru is near Jaaxali. We left Gunjuru and Jaaxali to migrate here. There was fighting; that is why we left.

After a quarrel among princes, the Bacili left Soxolo and went to rule in Banamba. After another quarrel, they went to Keniyu. Once settled there, they ruled from Keniyu to Gande. The Bacili spread their wing from Keniyu to Arundu, from Arundu to Heli Anma Dunbe Seruka, and over Bundu as well.

Al-Haji Maaliki Si came from Seeyunma, near Podor; his teacher sent him in search of grain. When he took the grain back and told his teacher where he had found it, the teacher said: 'Return to Bundu; God will make of it a robe for you to wear.' Alimaami Maaliki went to see the Bacili Tunka, and said he was a *moodi* looking for land to farm. The Tunka said: 'Come back on such and such a day; wherever we meet will be the boundary of the land I'll give you to farm.' On the appointed day the *moodi* set out in the middle of the night; the Soninke waited until they had eaten their morning *fonde* before setting out. They met with Si a short way beyond the *kolangal* floodlands. The Bacili was a *tunkan-yugo*, he kept his word; that was how the *Sisiganko* gained control of Bundu.

Our ancestor, named Lamiina Siliman, married four women: Siliman's mother, Lasana's mother, Musa's mother, and Xulle Tanja's mother. We are all equals; the eldest rules.

Lasana's mother died, and he was brought up with Siliman. Musa's mother died, and he was brought up with Xulle Tanja. They separated; those who were reared by the same mother stayed together. But we are all sons of one father. No one is better than another.

Al-Haji Abdullayi Tanbadu

Lahaseneni, the descendants of Lasana settled in Jaala, then in Gamera. There are only two households here descended from Lasana. But the descendants of Siliman, Xulle Tanja, and Sallifo Sanba came here. The first settled in the neighbourhood called Silimana. Sallifo Sanba's descendants belong to Xoje, Al-Haji Hamidu's household. Xulle Tanja's descendants founded the neighbourhood of Tanjankunda. Among them was Al-Xali Madi Bidan. He was one of the first to come; he was a learned man.

Al-Haji Jaaje Jaaxo

O ri taaxu wande deben ya; when we came, we found the Geyi and So here. We did not obtain land with guns. There is no one here whose ancestors fought for land. If you asked for land and had people to farm it, they would say: 'It's yours.' They couldn't farm it themselves, and they wouldn't sell it. Besides, any founder of a town wants people to come and make it grow.

Haaja Sedinte Bacili

Bacilinin do Jaaxalinko, the Bacili and the people of Jaaxali are children of the same mother, but different fathers. They are *moodini*; they read. We did not read.

The Tanjigoora, the people of Jaaxali, don't give us their daughters in marriage. No Bacili will ever marry a Jaaxali woman. They just don't, that's all. But they marry Bacili women; am I not here? They don't ask for their hand, they don't give bride-price; they take the women they are given in alms.

It was I who rejected the Bacili; I myself. My father didn't give me in alms! When my first husband died, many Bacili wanted to marry me, but I refused. I said: 'You Bacili, I don't like you. If I find a *moodi*, I'll give myself in alms.' So I gave myself to a *moodi*. When I told my father, he said: 'Eh! if you love him, I don't mind. Let them give you to the *moodi*!'

With us, when a child is born, you carry the child to the river to show it to the *muno*, a spirit that lives in the river. You say to the *muno*: 'Here is your great-grandchild.' The *moodi* don't like that. But it's what we do at home. I carried my grandchildren to the river. I said: 'Leave me alone, all of you. These children are related to me. It doesn't stop you being a good Muslim. *Ke ni Hadama*; it's the old Adam, it's a way of doing things. You love Islam and you love your country. It doesn't do any harm.

Mammadu Abudu Tanjigoora (Isa Budu)

Moodi do tunkan-yugo, cleric and prince, we're all sons of the same father and the same mother. One has chosen God; the other, the world. That's what it means to be a ruler. The *moodi* has chosen the book. He'll say yes to whatever he's told. The other has chosen the gun; if he sees someone to kill, he'll kill him; if he doesn't kill him, he'll make him a slave. He's the one who says, when he sees a country: 'This country is mine.' That's how he has land. If someone wants to come and settle on his land, he'll say: 'All right, come and settle.' He's the ruler. The *moodi* come and settle on his land. We follow God. That's the difference: some say 'God', and others say 'the world'.

My ancestor passed by here and went on to Tiyaabu. All that land belonged to the Bacili. He said he was looking for somewhere to settle. They said: 'Look at the country and choose wherever you like.' He went to Gallaade; it was good there, but too far. He came back, and liked it here. He found the people of Geyga here, and told them he wanted to settle. They said: 'Welcome; we are fishermen.' He settled right here, where our house is now. In those days there was nothing here: just trees and grass, not a single dwelling.

Our land by the river was given to our ancestor by the *debegume*. My ancestor's name was Al-Xaali Madi Bidan. The name of the *debegume* was Muulu Geyi. That was a long time ago.

Adrian Adams

It is written that Gunjuru and neighbouring towns were sacked in 1724 and again in 1730. The earliest wave of settlement from eastern

Gajaaga was certainly completed during the first half of the eighteenth century, for a map dated 1751 shows 'Cougran' at the right place on the south bank of the river, upstream from 'Segamani'.

In stories of this time there is scarcely any mention of a European presence. Yet the French were sending ships upriver from Saint-Louis in Buleeli Denba's day. It was Frenchmen, and a passing Scot on his way to the Niger, who noted half-heard place-names on maps of the river, and recorded details of the shifting alliances between the Bacili and their neighbours. In the late seventeenth and early eighteenth centuries, the clerics of eastern Gajaaga were purveyors of slaves to the Saint-Louis traders; some of the strife between the Bacili was caused by the French trading posts and the duty they paid.

The trade waned with the loss of Saint-Louis to the English in the mid-eighteenth century, and was abolished in principle before the French regained it in 1818. From 1818 to 1830, the French made unsuccessful efforts to establish an agricultural colony in the lower valley. In 1818 they chose the town of Bakel, in Gajaaga, as a forward base; a fort was built there, on a hill overlooking the river. But the renewal of River trade they hoped for was not to be.

Baraka

Jaabe So

N wa Fanqannen jingun tu, I know the boundaries of the land by the river because I farmed there as a boy. We grew up there; I know the boundaries of all the fields between here and Golomi, here and Bakel. People brought seedlings of *gundunbuse* and *xaamere* from the bush, and planted them to mark the edge of a field.

Also, my great-grandfather Baraka wrote it all down: the land belonging to the town, how people settled here, the boundaries of people's fields; what we call a *taarixu*. I've read his book. It was once in my possession.

Al-Haji MaJaaxo Mba Jaaxo

Ke ni taarixun ya de, *taarixu* are what they write on paper. In Arab countries they've been writing them for a long time. Writing is a gift of God; but we black people have neglected it.

Isa Budu

N rawa telle, I can go back as far as Al-Xaali Ancuman. Between Al-Xaali Madi Bidan and Al-Xaali Ancuman, this household had seven

The Saracolets are all Bakely or Marabou. A Bakely is a king, that is to say a totally independent man against whom there is no recourse in justice, whatever crime he may commit. They rob, plunder, and kill with impunity. The Marabous are those who trade. They travel every year, either to purchase captifs themselves or to bring traders from Bambarenas.... Although independent, the Saracolets have a king whose only right in the country is that of receiving the annual custom we pay, amounting to some 2,000 livres. He takes precedence over the other Bakelis without having any power over them, nor indeed any more power than they in any matter whatsoever. Although this king is elected, he must always be in turn the eldest Bakely of one of the four départements into which their country is divided, namely Tuabo, Lanel, Makrana and Caignoux, who is elected king at the death of him who wears the crown....

When we arrived in Tuabo, one of the most important towns of the country, I spent a whole day there in company of all the Bakelis who had gathered to question me about the purpose of my journey... I told them that as I had come to their country with the sole purpose of doing good... and had no intention of making any changes to the way trade had been carried out until now, I hoped that all the Bakelis would help me to establish not only the fort in Caignoux, but also any other I deemed necessary for increasing trade.

(Pierre David, Journal d'un voiage fait en Bambouc en 1744, ed. André Delcourt)

The kingdom of Kajaaga, in which I was now arrived, is called by the French Gallam; but the name that I have adopted is universally used by the natives. This country is bounded on the south-east and south by Bambouk; on the west by Bondou and Fouta Torra; and on the north by the river Senegal.

The air and climate are, I believe, more pure and salubrious than at any of the settlements towards the coast; the face of the country is everywhere interspersed with a pleasing variety of hills and valleys; and the windings of the Senegal river, which descends from the rocky hills of the interior, make the scenery on its banks very picturesque and beautiful.

The inhabitants are called Serawoollies, or (as the French write it) Seracolets... The government is monarchical; and the regal authority, from what I experienced of it, seems to be sufficiently formidable. The people themselves, however, complain of no oppression, and seemed all very anxious to support the king, in a contest he was going to enter into with the sovereign of Kasson. The Serawoollies are habitually a trading people; they formerly carried on a great commerce with the French in gold and slaves, and still maintain some traffic in slaves with the British factories on the Gambia. They are reckoned tolerably fair and just in their *dealings, but are indefatigable in their exertions to acquire wealth, and they derive considerable profits by the sale of salt and cotton cloth in distant countries....*

Dec. 25th.—About two o'clock in the morning a number of horsemen came into the town, and having awakened my landlord, talked to him for some time in the Serawoolli tongue. Madiboo informed me, that as they were dancing at Dramanet, ten horsemen belonging to Batcheri, king of the country, with his second son at their head, had arrived there, inquiring if the white man had passed... A short man, loaded with a remarkable number of saphies, opened the business in a very long harangue, informing me that I had entered the king's town without having first paid the duties, or given any present to the king, and that, according to the laws of the country, my people, cattle, and baggage were forfeited... They departed, having first robbed me of half my goods....

Toward evening, as I was sitting chewing straws, an old female slave passing by with a basket upon her head, asked me if I had got my dinner. As I thought she only laughed at me, I gave her no answer; but my boy told her that the king's people had robbed me of all my money. On hearing this, the good old woman immediately took the basket from her head, and showing me that it contained groundnuts, asked me if I could eat them.

(Mungo Park, Travels in Africa, 1795)

---

heads; but I don't know who they were. It must be written down somewhere. I'll have a look at my papers. But I lost many papers in the Tanjankunda fire, and my eyes are tired.

Jaabe So

Baraka d'i taarixun dabari, Baraka wrote his *taarixu* in Arabic. When he had finished, he called everyone to a meeting, one Friday evening at the time of *laxaasara* prayers. He said: 'Here is a *taarixu* that I have written. I myself am in it, and all my forebears; how everyone came here is also in it, and how everyone lives now. If anyone here was ever taken into slavery, and escaped, that too is in it. If anyone has done a good deed, it is here also. Read it, to see if it's true or not.'

So they read it through. There were debts recorded in it, and they said he shouldn't have put those in. He had a large herd, and when times were hard, people borrowed cattle from him, to sell. Those were the names he had written down; some said that was bad, because those people would be shamed if *faaba-renmu*, rivals, saw their names.

He said: 'No *faaba-renme* will ever see this. I've shown it to you because everything about Kuŋani is here: how people came here, what the land is like, what the people are like; those who are good to their slaves and those who are cruel to them; those who were themselves captured into slavery and then redeemed; those who had slaves and set them free in the name of God. How Guñan was founded is here, as well as Jaagili, Jogunturo, Yafera, and Golomi. I myself witnessed the founding of the latter two towns. It was my father who was summoned to welcome the people who settled in Yafera; and my father whom the Bacili sent to sell Jogunturo for a male and a female slave. I didn't include Gucube. Gucube was here before Kuŋani; it was the first of all the towns along this stretch of the river. It's all written down: look at it, and see if I've made any mistakes.'

He'd written down all the Bacili battles and where they were fought. He'd included the time when troops sent by the Bacili of Kammera camped on the north bank of the river, where the Jaagili school is now, and his father Buleeli Haaruna collected gold to pay them to leave: at dawn their chief said 'Nebe taxa, I am leaving.' He wrote it all down. He wrote some things down that no one could say aloud, even now. Nothing was left out. They all approved what he had written; then he put it away.

It was like a book: many sheets of folded paper, in a cover bound with cloth of local weaving, like the woman's wrapper they call *fendeli binne*; indigo-dyed, not black but dark blue. That book was once in my possession. I read it when I was young. But Salun Geyi borrowed it from me; he died, and the book was never found.

Adrian Adams

Two and a half miles along the river: there is no knowing how much land that was. So much has been lost to the river, that can never be reclaimed; the bank eaten away, the roots of trees laid bare before they fall. What is lost can only be known through what remains.

Founding ancestors are better remembered than those who came after them; more is known of Buleeli Denba than of his son by Seega, Haaruna. Haaruna is said to have farmed Fanqanne, growing mostly *gajaba* sorghum. He married a woman named Asiya Waaju, from Sobuku in eastern Gajaaga, and had a son known as Buleeli Haaruna or Buleeli Asiya. Buleeli Haaruna had several sons: the eldest was named Baraka. Haaruna, Buleeli Asiya, Baraka: it is because a little is remembered of their story that we can speak of this land's history. Baraka knew this. If it is now remembered that certain things happened in

By the River (c.1720–1933)

Haaruna or Buleeli Haaruna's day, it is because Baraka wrote that chronicle that is now lost.

In Buleeli Haaruna's day, a Gunjamu household came here from Gallaade downstream where they had first settled. The Gunjamu were *mangu*, warrior vassals of the Bacili. They were given rough land to clear, upstream from the land then under cultivation at Fanqanne. Two Jabira households from north of the river, *tunka-lenmu* who settled here in Baraka's day, were given land just downstream from the Gunjamu. The furthest upstream of all Kuŋani's land was given to the Tanbadu, clerics from the Soninke state of Jaafunu, then under Bambara domination.

The new towns founded in Haaruna's day, Jaagili just across the river, Bakel five miles downstream, Golomi three miles upstream, were founded by people not from the east, but from north of the river or further downstream; most of them *mangu* or *tunka-lenmu* come to take part in the Bacili wars, like those who in Buleeli Haaruna's day founded Yafera, ten miles upstream from here, and Jogunturo on the north bank, just opposite where the *Fan-lenme* flows into the *Fan-xoore*.

The Fula state of Bundu, founded at the end of the seventeenth century, encroached upon Gajaaga from the south; as did Fuuta Tooro from the west, Xaaso and Kaarta from the east, and bands of 'Moroccans' from the north. But the fighting was not so much against outsiders as between warring Bacili factions, using mercenaries. These conflicts, endemic since Buleeli Denba's time, flared into civil war in Baraka's day. He, like the recently settled *mangu*, was on the side of the Tiyaabu Bacili, against Maxaana.

No Bacili household ever settled in Kuŋani; those who came from the east to found Tiyaabu, stopped here briefly, then moved on. As the one clerical town of western Gajaaga, it enjoyed neutrality: the *moodini* did not pay tribute, nor take part in wars. The town was beginning to gain a reputation for Koranic studies, centred about the Tanjigoora household of Xoje.

Jaabe So

I ga ri, when the Gunjamu first came, they went on to settle in Gallaade. One Gunjamu household returned here; that was in Bulel Asiya's day. They asked Geyga for land. Geyga gave them land quite far upstream, saying: 'If you can clear the *kenŋe* grass from the land, go and clear it.' In those days, Moors would lie in wait for people in the bush. The Gunjamu had many slaves: they cleared the *kenŋe* and began farming the land.

All our grandfathers, our great-grandfathers, our remote ancestors were united: the Bacili, their *mangu*, their *sakko*. The Bacili never went to war without the Gunjamu. They never went to war without the Jallo, the *mangu* of Manayeli. The Timmera of Yafera are *mangu*, but when they arrived, the wars were over. The *moodini* are like women.

When the Bacili ruled, we were their organizers, all along the river. God did not create the world for one person; but there are distinctions that remain alive. We are powerless; we are not rich. But we have ways that we cannot lose. The Bacili have ways that they cannot lose. The *mangu* have ways that they cannot lose.

Who can make us leave Tunka Sanba's chamber? The Bacili called us *Manga-Sakke*. We tied a strip of red cloth round Manga's horns, and lowered him alive into the grave; we stood a living person alongside, to be buried with him. Who can sever us from the Bacili?

Al-Haji Jaaje Jaaxo

O do ñiiñe ma ri, we did not come here with land, nor with guns; we came with books. We were *moodini* from the beginning. The task of every person named Tanjigoora is to read.

At one time, the Tunka of Maxaana was called Sanba Yaasin; the Tunka of Tiyaabu was called Sanba Xunba Jaaman. They fought each other.

Sanba Yaasin destroyed Tiyaabu; he killed a hundred first-born sons. Sanba Xunba Jaaman came here to Kuŋani, and took refuge with a great *moodi* whose name was Amara Jaaxo. The Tunka of Maxaana asked: 'Have you brought me my namesake's head?' When they said no, he said: 'Tiyaabu has not been destroyed.' He said they should post men in Golomi, Yafera, Jaagili, Gaabu, Bakel. When Sanba Xunba Jaaman heard this, he would come out, and they would kill him.

Suraqe Faatuma Jaara, of Laani, said: 'No, that must not happen. We must not fight where there are women.' The Tunka said: 'Are you setting yourself against me?' Suraqe Fatuma Jaara said: 'Did you give me your daughter, that I might not oppose you in public? No woman can keep me from telling the truth. I'm no traitor. I will go and speak to Sanba Xunba Jaaman in Kuŋani.'

He found him with Amara Jaaxo, and said: 'Bacili, if ever I have the chance, I'll kill you. If you ever have the chance, you'll kill me. But I don't want you to be shamed. Sanba Yaasin has laid a trap for you.'

Sanba Xunba Jaaman said: 'I understand. Be thanked; you have behaved honourably. Whoever does not want to see me shamed, is a friend.'

Amara Jaaxo took his cap and set it on Sanba Xunba Jaaman's head. The Tunka said: 'Where shall I go?' The *moodi* said: 'Wherever you will.' He went north, to the Moors' country. When he came back, he paid his score by destroying Maxaana. He killed 160 men.

The *tubab* came and made peace between them. That was the time when the *tubab* were beginning to have power. The two Tunka were reconciled at Jondisaaye, between Yafera and Golomi. Afterwards, each ruled his own country.

Jaabe So

Barakan xoxone, Baraka's younger brother Amara went to work for the *tubab* in Ndar, as a gardener. The daughter of a Fula government employee fell in love with Amara. She cast a spell upon him to make him ask for her hand in marriage. At that time he was already engaged to a *sakke* girl in Manayeli, Menti Gaajigo; but he did not marry her.

His brother went after him, but found that his wife was pregnant. Amara wept. He said: '*I taata, nke ña gundunke*. O my brother, I have become a person of the bush.'

I was once told a story about my great-grandfather Baraka. One day some children were running and singing a song that went: '*Toqqo toqqo falle, maama yin-kanba*: stay, stay behind, eat your grandmother's head.' People laughed and said: 'Do you hear what those children are singing?'

Baraka said: 'The song has a hidden meaning. In youth you live with your brothers and your wives, all of you young together. If God gives you long life, they will all die and leave you alone with the children, thinking and talking to yourself. The children's song is about the person who is left behind. Everyone has died and left you all alone. You eat your grandmother's head; that is, you talk to yourself, and no one understands what you say.'

Mammadu Denba

Adrian Adams

After their return to Saint-Louis, the French moved upriver with a view to establishing direct control. In 1818 they built a fort in Bakel, on a hill by the river. Al-Haji Umar tried to expel them, but failed: the treaty he signed in 1857 acknowledged French dominion as far upstream as the confluent with the *Fan-lenme*.

There were reprisals against Gajaaga towns. The protectorate treaty of 1858 established French rule over Bakel and the part of Gwey upstream from Bakel—Kuŋani, Golomi, Yafera, Arundu, and Baalu.

Umarian rule to the east, and the progress of French conquest, blocked inland trade from the mid-nineteenth century. In 1887 all of Gwey, with Kammera upstream and Gidimaxa on the north bank, came under French control; by the end of the century, railway-building and the growth of groundnut farming near the coast turned their attention away from the River.

Jaabe So

N faaba, my father was Amma So; Amma So's father was Mammadu Denba. Mammadu Denba and Usumaanu were sons of the same father, Baraka. Usumaanu was called *Sakke-boroke*, 'the sterile *sakke*'. When Baraka died, Mammadu Denba was still young; his mother took him back to Xaayi with her, since there was no one here to marry her. When *Sakke-boroke* died, Konko Goola Bacili's father and Madi Moodi Geyi's father got together and decided they should go and look for Mammadu Denba; otherwise their *sakke* household would be at an end. They went, and brought him back to Kuŋani.

The Bacili and the people of Geyga brought him back to Kuŋani; the Bacili asked to take him on to Tiyaabu. The people of Tiyaabu honoured him and made him welcome, as a person lost and found again. They gave him a wife, Mariyam Siise from Bakel. As bride-price, the Bacili gave a *falo* here in Kuŋani, and the land above it: at Sanba-Salu, next to the land given to the Tanjigoora household known as *Bundunko*. Two of Mammadu Denba's children lived to adulthood: my father Amma So, and his sister Munune, who married in Sobuku.

Mammadu Denba was with the Bacili, not as a courtier, but as an adviser, an organizer. If he said something was bad, they would leave it alone.

In those days it was still the custom for the Tunka to spend the first night of his reign with a *sakke* maiden. From such unions only daughters were born. It is known even now that the men of valour in this or that household are descended from these *sakke* women. But Mammadu Denba put an end to the custom, because he saw that in times ahead there would be no unity.

Mammadu Denba did not farm. He had a large herd of cattle, and received grain from Tiyaabu every year at the end of the rainy season. His slaves didn't farm for him either; he gave them land to farm for themselves. One woman, Faatuma Denba, used to farm the *falo*. People would say: 'All that land for one woman slave!' She would laugh and say: 'When the master cooks meat, the slave eats it.'

Mammadu Denba treated his slaves like brothers; they lived as members of one household. The slaves herded cattle; he fed and clothed them, and sought wives for them. People said: 'Mammadu Denba is spoiling the slaves; he has land, and he doesn't make them work it.' When Mammadu Denba heard this, he said: 'I didn't capture these slaves, nor did I buy them. The Bacili gave them to me. I consider them my children.' He always refused to take a *taara*, a slave concubine.

Some strangers came from Gambia to see Seexu Amara: Haaruna So, his son Moodi, and his daughter Juumu. Seexu Amara entrusted them to Mammadu Denba. Mammadu Denba gave his daughter in marriage to Moodi So, and his son Amma married Juumu.

Adrian Adams

The Geyi were not many in number, nor were the descendants of Buleeli Denba So. They gave away the last of the land in the further reaches of Fanqanne: to Xoje, which remained the paramount household for Koranic teaching; to other Tanjigoora household heads, often related to the Geyi by marriage, like Buna Fasunte of Tanjankunda, a descendant of Al-Xaali Madi Bidan, whose mother was a Geyi; and in one instance, to the daughter of a runaway female slave.

Plots from their own reserve of land near the town were given, then and later, to households descended from those who had already received grants of land more distant from the town; occasionally to households that had no other land at Fanqanne. Their reserves dwindled further as the river bank was eaten away by the river.

In Mammadu Denba's day, there was no more land at Fanqanne for new Tanjigoora arrivals, the household known as *Bundunko*, 'the people of Bundu', or that of Foode Mammadu Arjana. They cleared plots inland from Fanqanne, or were given larger tracts of riverside land to clear downstream, on the other side of the town from Fanqanne, at a place known as Sanba-Salu, heavily wooded and subject to flooding. Later grants of land at Sanba-Salu were also made to horsemen of note, like Sanba Haawa and Muusa Kaba. At Sanba-Salu, the Geyi kept some land near the town for themselves; and the So, a large plot a mile from the town.

The town also grew. Early households divided, one branch remaining in the old compound while others, like Buna Fasunte and Sanba Haawa, built on a new site. New arrivals settled in Silimana. Mammadu Denba moved the So household down from Geyga to a place just downstream from Xoje's first site; and Xoje moved up to Geyga, where the So lived before. The place at the foot of Geyga hill where the So cattle were kept was also given to Xoje by Mammadu Denba, to build a mosque.

Amara-*xoore*, Amara the Elder, had died by then, and was succeeded at the head of Xoje by his brother's son, Amara-*tugune* or Amara the Younger, known as Seexu Jomo. It was in Seexu Jomo's time, and Mammadu Denba's, that Al-Haji Umar Tal, a Muslim leader from the Pulaar-speaking Middle Valley, having failed to evict the French from the River, passed through Gajaaga at the head of an exodus towards the east, away from a land ruled by infidels. Few people left Gajaaga to join him, although they had supported him earlier. But thirty years later, Mammadu Denba, his son Amma So, and Seexu Jomo all supported the Soninke cleric from Gunjuru, Seexu Mammadu Lamiina Daraame.

Al-Haji Mpali Tanjigoora

Our first person here was my great-grandfather, Fode Madu Buukari. He lived in Gambia before coming here. He begat Foode Bakari, Foode Bakari Jaaxo. Foode Bakari Jaaxo begat our fathers.

It was by hard work that they won our land at Fanqanne. When they came here, they had no land. One day after *laxaasara* prayers, my grandfather Foode Bakari went to see the *debegume*. He said he wanted the land by the burial ground, on the side nearest the river. It had once been farmed, but at that time lay abandoned; it was so overgrown that panthers would hide there to kill calves. When our grandfather spoke to the *debegume*, he replied: 'Yes, but that land belongs to someone. Wait until tomorrow after *laxaasara*. I'll tell them: "You know, that land near the burial ground is so overgrown that wild animals hide there and kill our cattle. If those it belongs to cannot farm it, I'll give it to someone else."'

They agreed to that. My grandfather kept quiet about it. The next day, after the *debegume* had spoken, people said: 'We cannot farm that land.' At that time, that land was covered with a kind of grass called kenŋe. The *debegume* went to Foode Bakari and said: 'Come and take the land you want; you're free to clear it.' Foode Bakari had a large household then; they cleared all the land he had been given. It lies between Malle Joogu's field by the river, and the pond near the burial ground. They started near the *debegume*'s land at Munune, that was later given to Xoje; they cleared past the Daraame field, up to the land that belongs to people from Tanjankunda. Then Foode Bakari went back to see the *debegume*, and said: 'Well, I have a field now, but my slaves have no land to farm for themselves. Please help me out again.' The *debegume* agreed, and showed him a place. So the slaves had land to farm, the other side of the Si household's field from the first piece of land. Our household was strong in those days; that land was won by their strength.

According to what our fathers told us, our land at Guñaña, between here and Bakel, was also won by hard work. At that time, Moors would lie in wait between here and Bakel. No one could go to Bakel; if you went, they'd kill you. That was Bacili land, but our grandfather received it through the *debegume*. The household to which Muusa Kaba belonged was given land next to theirs. When they cleared it, one group of people would fell trees, while another group, armed, kept watch in the trees. Life was hard in those days.

Jaabe So

When the *tubab* came here, they fired their guns all the way from Saint-Louis to Xaayi. When they came to Manayeli, they fired their guns, but they didn't enter the town.

When they found they couldn't defeat the *mangu*, they said: 'Let's go on to Tiyaabu.' They went by the inland route, past the hill, to attack Tiyaabu. The people of Manayeli said: 'Let's go and help the people of Tiyaabu; otherwise, from what we've seen of those *tubab*, they may defeat them.' They went to Tiyaabu and helped fight the *tubab*, for three days. The *tubab* did not enter Tiyaabu. They went on to Bakel.

The *tubab* never entered those two towns. It's been put in writing; the paper is in Bakel. The *tubab* ruled all the rest of the country. But Tiyaabu and Manayeli remained independent towns within the kingdom of Gajaaga. My grandfather Mammadu Denba took part in those battles. That was the first fighting there was between the Bacili and the *tubab*, long before Seexu Mammadu Lamiina.

Tiyaabu and Manayeli were like right hand and left: each knows what the other is doing.

Al-Haji Sanba Jimmera

The school of Koranic studies at Xoje was founded by Amara Jaaxo the Elder. It was he who brought up Seexu Jomo, who later supported Seexu Mammadu Lamiina.

Al-Haji Jaaje Jaaxo

When Amara the Elder died, Amara the Younger (also called Seexu Jomo, Jomo being his mother's name) had not yet finished his studies. It was Foode Salli Jaaxo, from Al-Xaali Jaaxo's household, who completed his education.

Al-Haji Umaaru and his people from Fuuta and Bundu came to see the Tanjigoora. He was a Muslim; we gave him food. But no one left here to follow him.

Jaabe So

Seex Umaaru passed by here. He first went to Kidira; he cursed Kidira, saying: 'May this town never know good fortune.' He spent the night in Jaabalu, and invoked blessings upon them. He went to recruit followers in Arundu. Some people from Arundu joined him, but no one from Kuŋani.

That was in Seexu Amara Jaaxo's day. Al-Haji Umaaru met with Seexu Amara Jaaxo; they spoke at length, and reached an understanding.

Sanba Si

Seex Umaaru came to Tiyaabu. He asked people to join the jihad; the Bacili said no. They refused. He went to Gaabu, and raised some people there.

In those days, Seexu Jomo was living here; he was a *waliyu*, a wise man. That's why Seex Umaaru couldn't find any followers here.

No one saw him die. No one could say he saw Seex Umaaru's body. He fought the jihad in forty-four towns; he devoted himself to God's work.

I was born in Mbumba, in Fuuta. I am an old man. I have seen people who saw Seex Umaaru.

Al-Haji Jaaje Jaaxo

It was Sanba Xunba Jaaman's son Silli Fanna who sold Bakel to Faidherbe for a hundred pieces of silver, along with our *cercle* and our burial grounds; for a hundred pieces of silver.

During the first months of the year 1855 the warriors of Guoye, gathered in the Tiyaabu fort, had warded off an attack by the volunteers from Bakel. In August of the same year, a column 450 men strong, including 250 marine infantrymen, attacked the village again in an attempt to avenge this defeat. But the expedition failed; the column was obliged to beat a retreat to the avisos, leaving behind 10 dead and bringing back 51 wounded.

(Colonel H. Frey, Campagne dans le Haut Sénégal et dans le Haut Niger, 1885–86, 1888)

In July 1855, Captain Parent went to burn down the village of Counguel (sic) the largest village of Guoy after Tuabo... In May 1858, men from the fort in Bakel destroyed the village of Kounguel (sic), where hostile elements from Diaguili had taken refuge.

(L. Faidherbe, Le Sénégal, 1889)

From 1855 to 1858, during the colony's enforced struggle against El-Hadj Omar, the Sarrakholais, moved by Muslim fanaticism as much as by love of pillage, came out in favour of the new prophet and for a long time remained his most faithful partisans... Then, ruined by El-Hadj Omar's war, decimated by the dreadful famine that ensued, the Sarrakholais made their submission.

(Frey, Campagne)

In Kounghani, at the time Oumar passed by there, there was a very learned grand marabout named Cheikh Ndiamo Tandjigora. With such a marabout, people could do without Oumar.

(Sall interview, Fonds Robinson, Institut Fondamental d'Afrique Noire)

Seexu Mammadu Lamiina studied in Bakel. His family name was Daraame. He studied under Foode Bakari Daraame, who lived for a hundred and thirty years. When he had finished his studies, he went on the pilgrimage to Mecca. On his way back, he went to Seegu. At that time, LaamDo JuulBe was ruler there; when LaamDo JuulBe saw how great his fame was, he detained him.

Seexu Mammadu Lamiina did not go to prayers in the mosque. When LaamDo JuulBe challenged him, he said: 'I won't go; if you kill me, I will meet with the Prophet, and you with Iblis.' When LaamDo JuulBe asked why he said that, he said: 'Where are the Prophet's disciples?' The other answered: 'They're dead.' He asked:

'Where is your father?' The other was silent. He said: 'Your father's dead.'

Seexu Mammadu Lamiina married a woman named Barakata; to trick LaamDo JuulBe, until he could escape his dominion and go his own way. At that time he intended to lead a jihad in Gamo. But he met up with Bokar Saada and Umaar Gawula, and their double-dealing led to trouble.

Isa Budu

Tibaabu ga ri, when the Europeans came here, Seexu Mammadu Lamiina arrived from the East. He said: 'Let us unite to expel the infidels from the land. Infidels should not settle here; this land is for Islam alone. Let us unite and fight; let us be true Muslims, and study the Koran. Let us set aside what God has forbidden, and do what he has prescribed. We must not let infidels rule over us.'

Everyone said: 'Yes, that's true.' He said: 'Now let us wage a jihad.' They agreed.

Jaabe So

Seexu Mammadu Lamiina went to see the Bacili. He told them he had come to wage war, and needed someone to help him organize. The Bacili told him to go and see Mammadu Denba in Kuŋani.

Seexu Mammadu Lamiina came here at night, and knocked at Mammadu Denba's door. Mammadu Denba bade him welcome. Seexu Mammadu Lamiina told him that he intended to fight the *tubab*, and asked him to make this known throughout the land and enlist people's help. Mammadu Denba said that he would first take him to meet Seexu Amara Jaaxo. That night, Seexu Mammadu Lamiina and Seexu Amara Jaaxo conversed until the stars came out. Then Mammadu Denba and Seexu Amara Jaaxo escorted Seexu Mammadu Lamiina and his four *taalibo* out of the town on horseback. Mammadu Denba took the inland route through Bema, and entered Bakel at daybreak.

Mammadu Denba set to work. He travelled on horseback from here to Bundu; they called him *si-binnen gunme*, the man with the black horse. There wasn't a town in Bundu that didn't hear from him: 'Seexu Mammadu Lamiina has come to wage war; every man of worth must help him. His troops are in Kuŋani.'

He went into Bundu and came out again. He went into Hayire and came out again. He went to Gidimaxa: Testaayi, Teyesiibi, Jaajibinne.

By the River (c.1720–1933)

It was Mammadu Denba So who organized Seexu Mammadu Lamiina's war. People talk of Seexu Mammadu Lamiina's war, but they don't know who organized it. Mammadu Denba begat Amma So, and Amma So begat Jaabe So. Mammadu Denba's father was Baraka. All *sakko* from Kuŋani.

Mammadu Denba would go to Tiyaabu through Bundu, change into dark clothes and enter Bakel to find out about preparations for war. Then, in the dead of night, he would saddle his horse and leave the town. He would tie up his horse near one of the Fula towns in Bundu; in the morning he would ride back to report to Seexu Mammadu Lamiina. He was Seexu Mammadu Lamiina's spy.

Isa Budu

Seexu Mammadu Lamiina went as far as Xaayi and Gidimaxa, and he roused them all. All the people set out for the infidels' country near Velingara, above Tambacounda. They came to the Fula country, Bundu. They said they wanted to pass through there. The Fula said: 'Where are you going?' They said they were going to the infidels' country. The Fula said: 'Fine, but don't come through here.' They said: 'Yes, we will.' They fought the Fula, and spoilt the water of the country's wells; they killed them all.

Sanba Si

Seexu Mammadu Lamiina's war crossed the *Fan-lenme* to Bulebone. In those days, the ruler there was Umaar Penda, of the Si family. Seexu Mammadu Lamiina said he wanted to pass through his country; he said no.

Jaabe So

Bulebone was the first battle of Seexu Mammadu Lamiina's war. If someone says he's going to tell you about that battle, that happened over a hundred years ago, most likely that person didn't witness it himself. My aunt Juumu Haaruna, my father's first wife, told us about it; so did my father's friend Sanba Haawa.

Those who took part in the attack on Bulebone were from Jaagili, Golomi, Yafera, Arundu, and Kuŋani. Bulebone was in the kingdom of Xaaso. It was surrounded by a high wall, like all rulers' strongholds in the old days, and was very difficult to enter.

The people of Bulebone didn't come out; they stayed behind the wall, which had rifle-holes in it. The men I've heard tell were brave that day, were Sanba Haawa, Muusa Kaba, a Jabira from Jaagili and two *Koliyaajo* from Golomi. But Sanba Haawa was the bravest of them all. My father was Sanba's friend, and was there to help him.

At the start of the battle, with all Xaaso inside the wall, people said: 'Who will be man enough to climb it?' Some went into the bush and cut poles to make ladders. Everyone was making ladders. My father said to Sanba: 'Put your gun down; lean it against the wall.' He did. My father was a strong man, and tall. He knelt down and Sanba stood on his shoulders; then he said: 'Hold on to the wall' and he stood up. Sanba took hold and pulled himself up to sit on top of the wall. Then my father handed Sanba his gun, and he began to shoot.

The people of Bulebone had a fetish that day, a pregnant woman seated naked on a new calabash. Sanba shot her first; then he leapt down with his gun, and started the battle. Sanba was the first to be brave that day. The people who had made ladders climbed them to join him. That was how Gajaaga defeated Bulebone, in the country of Xaaso.

Isa Budu

The survivors fled and took refuge with the *tubab*. The others said to the *tubab*: 'Hand over those people.' The *tubab* said no.

When the *tubab* first came here, the Bacili, who owned the land, said: 'Welcome to Bakel.' They said: 'We are traders; we will bring goods to sell.' The Bacili said: 'That's good. You're not here to fight, nor to do any harm; you have come to trade. You are welcome.'

They didn't fight anyone. But the Fula fought, then fled to the *tubab* in Bakel; and the *tubab* refused to hand them over. So the country fought the *tubab*.

Fula, *tubab*—the country fought them. The Bacili who were for the *tubab* fought on their side. The Bacili who were for Islam supported Islam. The war lasted a long time. Seexu Mammadu Lamiina waged war against the *tubab* because they broke their promise; they came to stay, to ruin our land. Many people supported him.

Sanba Si

He had all Gajaaga with him; everyone followed the *moodi*. Men like Sanba Haawa, Muusa Kaba: they all followed the *moodi*. They decided to go to Bakel. When they came to Sanba-Salu, between here and Bakel, they halted there.

Maama Kaba Tanjigoora

My father Muusa Kaba was here at the time of the battle of Sanba-Salu. They said: 'Tanjankunda and Silimana, there will be deaths tomorrow! Forward to Sanba-Salu; there will be deaths tomorrow! Let no one who is not sure of himself go to battle tomorrow.' My father said: 'If I'm not first in battle, I won't be the last.'

There was drumming that day for the warriors, in the Konate household; my father's sisters were there. There was fighting at Sanba-Salu, and people were carried dying into town. The *tubab* who said he would destroy Gajaaga was loading the cannon. My father prepared to shoot him. Just then, a Tanjigoora came up behind my father in order to kill him.

It was as if something told my father to turn around. He saw this man creeping up behind him, and said: 'If you don't stop where you are, I'll knock you over.' Then my father shot the *tubab* at the cannon, just as he was about to fire.

Jaabe So

The main battle of Seexu Mammadu Lamiina's war was fought on the banks of the dry stream-bed between Sanba-Salu and Seegankaani: among the rocks and trees there.

Seexu Mammadu Lamiina's people had no field-glasses to watch for the troops from Bakel. Instead, they climbed tall trees, *jebe*, *saye*, to keep a look-out. The fighters all gathered among the tall trees at Sanba-Salu; they waited for the French to arrive.

At the first exchange of fire, a bullet struck Sanba Haawa's brother, same father, same mother. People said: 'Sanba, your brother's been shot. Don't let them strip him!' In those days, if you shot a man, you would strip him and take his amulets. But no one did that to Sanba's brother. Sanba knelt astride him and kept on firing at the troops. Amma So was beside Sanba, loading his rifle and handing to him, then loading his brother's rifle ready to hand. He kept on shooting until the troops retreated.

The fighting continued. It was Muusa Kaba who thrust *xaame* leaves into the mouth of the cannon, killed the soldier manning it, and captured the cannon itself. That cannon is now in the mosque at Jaagili. They pursued the troops all the way to Bakel. But they found that the troops had taken refuge in the fort that is now the *préfecture*.

> *A large band of Sarrakholais, returning from a raid in Boundou where they had sacked several villages, established itself in Kounghel, a village located six kilometres from Bakel, on the way to Kayes. This band proposed to act in the same way towards our allied villages in the Bakel area.*
>
> *Captain Lefranc had ordered the* marabout's *partisans gathered in Kounghel to disperse, or else be forced to do so. The time-limit they were given had expired. [On 14 March, 1886], the* commandant *in Bakel ... sent the 1st company of* tirailleurs *to hasten their departure. ...*
>
> *The company crossed the creek-bed and entered a winding path, bordered by tall grass that impeded vision. ... It was attacked by a swarm of Blacks, who crept up upon them unseen under cover of brushwood and tall grass, until they were thirty metres from the troops. Many* tirailleurs *were killed or wounded; the porters threw down their loads and fled. Maréchal des logis* Besnier, *manning the cannon, was struck by two bullets in the thigh; sergeants Mariani and Samba Dieri were put out of action; MM. Laty and Toumané were wounded. ...*
>
> *The enemy's losses must have been very high. The battle began at five in the morning, and was not over until nine.*
>
> (Frey, Campagne)

They had shut the iron gates; no one could come out, and no one could get in.

Sanba and Amma So brought Sanba's brother here, then set out to take him to Jawara. In those days there was a healer in Jawara who could extract bullets from wounds. Alas! He died on the way. There's a tall tree by an anthill between Yelingara and Jawara, near Sanba Daraame's garden. Sanba Haawa's brother is buried there. Even today, when people pass by there, they say: 'That's the grave of Sanba *bu* Ali's brother.' That's a praise-name for a brave horseman; instead of saying 'Sanba', they say 'Sanba *bu* Ali'.

Sanba said: 'I'm going to Bakel with my gun, to attack the *tubab*.' Amma So said: 'No, Sanba. Let's cross the river into Gidimaxa and return to Kuŋani that way.'

That is what they did. Sanba was a brave man. Amma So was like a father to him, and did all he could to help him achieve honour.

Isa Budu

The people in the fort had to do something. They heated water; they heated and heated it. They poured the water. *Worrr—ssasssa—a teye!* It was hot! Everyone left the walls. Very hot! They had killed the men

> List of the most perfidious persons of the Bakel treachery... Fodéa Diam (marabout and chief trouble-maker, Kounguel).
>
> (Archives du Sénégal, 13 G 240)

> **5h05** Elingara, *left bank. These* Sarakolet *villages are really pretty, with their slender date-palms and huge baobabs...*
> The captain tells me that from Tenabo [sic] on, there will be fewer palm-trees, as the column of 86 [NB: Galliéni's column against Mamadou Lamine] completely destroyed, he says, all the villages between Tenabo and Samé.
>
> (Charles Monteil, 'La Remontée du Sénégal en chaland de Saint-Louis à Médine, du 24 février au 20 mars 1897', Bulletin de l'Institut Fondamental d'Afrique Noire)

> Three thousand men were killed by the French during the six-week campaign against Mamadou Lamine.
>
> (L. Faidherbe, Le Sénégal, 1889)

manning the cannon, all of them. It was then that they heated the water. When they poured it, all those who were near the wall fled.

After that, the *tubab* came and destroyed Kuɲani. They set fire to it. They went to Jaagili and destroyed it too. All Jaagili fled into the bush. Everyone! There was no one left. If they saw anyone, they killed him. Everyone fled.

Al-Haji Abdullayi Tanbadu

After the battle of Sanba-Salu, everyone went away. That is when the Bambara of Bema came here. They wanted to settle somewhere with floodlands to farm.

Sanba Si

The *commandant* asked: 'Where is your *grand marabout*, Seexu Jomo?' Seexu Jomo had gone to Gori.

Al-Haji Abdullayi Tanbadu

At that time, Seexu Jomo was close to Seexu Mammadu Lamiina's son Siyayibu. Siyayibu said: 'I'm going home.' Seexu Jomo said: 'Don't go'; but Siyayibu didn't listen to him. When he arrived in Digokori, he was taken prisoner. The *tubab* were then in Gucube. They took Siyayibu to Gucube, and killed him there.

Sanba Si

Seexu Jomo went to Gori. LaamDo JuulBe, Seex Umar's son, told the people of Gori to pay tithes to him. They refused; they said they would pay tithes to the *moodi* who was living with them. LaamDo JuulBe said: 'Is that so? My father converted you to Islam, and now you disobey me!' He surrounded the town. He said: 'I won't attack you now, because Seexu Jomo is staying with you: a good man, a slave of God, who has had to flee the French *commandant*. But when he leaves, I will strike hard.'

Supplies ran out, and they began to starve. The people of Gori said: 'There is no more grain in our granaries.' Seexu Jomo said: '*Sallahu*, we'll leave tonight.' Sanba Dukure's grandfather, Salle Dukure, was then living in Gori. That night, while the people of Gori were asleep, the two men left the town and went to Gemu.

Al-Haji Abdullayi Tanbadu

In those days, Seexu Jomo had gone to Gori. When their grain ran out, he went on to Gemu. Then he stayed hidden in Fuuta for a time. He married there; Lasana Isa's wife Jenaba is the daughter of his son from that marriage.

Buna Fasunte came back to settle here, he went to ask the *commandant* for news of him. He was told that Seexu Jomo had been seen in battle. He said: 'No, he never fired a shot. His name was spoken; but he never fired a shot.' The *commandant* said: 'If he ever takes up arms again, we'll have your own head.' Buna Fasunte went to bring Seexu Jomo back from Fuuta; they returned here together.

Isa Budu

At that time there was nothing left here, not a single dwelling. Our ancestor Buna Fasunte came back. He went to see the *tubab* in Bakel. They said: 'Who are you?' He said: 'I am Buna Fasunte.' They said: 'What do you want?' He said: 'I've come back.' They said: 'No, the people of Bema and the Fula have said they want to buy your town. We're going to sell it to them.'

Buna Fasunte said: 'I'm going home.' They said: 'No; you all left, you ran away. And since you've come back, we're going to take you prisoner.' He said: 'No; the war's over now.' They said: 'All right, you can go home. We give you permission to do so.'

They also said: 'You'll have to buy your town back from us.' He said: 'How much are you asking?' They said: 'We'll sell it for such-and-such a price.' He paid the *tubab* the amount they had asked for; they were surprised, because it was a large sum. He paid them, and they made him *debegume*. Not because he was related to the Geyi, nor as a Tanjigoora, nothing like that; it wasn't for him to pass on to his children, it was just for him.

Al-Haji Abdullayi Tanbadu

All the townspeople gave money to get the Bambara to leave, so that they could come back and settle in the town once more. Buna Fasunte was sent to give the *tubab* the money. The Geyi who should have been *debegume* wasn't there at the time, and Buna Fasunte's mother was a Geyi. That's why he was allowed to be *debegume*.

Jaabe So

Buna Fasunte was temporarily made *debegume*; his mother was Modi Geyi's sister. My grandfather Mammadu Denba could not become *debegume*, because he was often away in the bush with his cattle, at Seegankani. But he still took part in the town's affairs. It was the townspeople who sent Buna Fasunte to bring Seexu Jomo home.

People began to realize that Buna Fasunte had plotted with the *tubab* to take over as *debegume*. Mammadu Denba called a meeting of the whole town, to say that Buna Fasunte should not be *debegume*. They fetched Modi Geyi's son home from Kaolack. After that, we decided never to entrust the position to an outsider. Even if there's only a woman left to take it, we'll respect her.

Isa Budu

The people of Geyga came back. Buna Fasunte told them: 'I didn't want to take your place. The *tubab* made me take it. Have it back.' That was when the people of Geyga said: 'Take Garsingide, it's yours.' That's how we came to have Garsingide. Garsingide is *jeeri* land, not riverside land; but it belongs to us. We are the only people to own *jeeri* land in that way. God is good.

Al-Haji Abdullayi Tanbadu

I was a child when Buna Fasunte died. Seexu Jomo died later on, when I was a youth; he was present at my coming-of-age ceremony.

Jaabe So

When the war was over, the *tubab* in Bakel killed many of the bravest men here. When they had done with killing, they summoned the *notables* to discuss taxes. My grandfather went, with Buna Fasunte. The *commandant* said that each person should pay a *muud*, a measure of grain. Buna Fasunte said: 'Perhaps three *nigife*, three quarter-measures.' Mammadu Denba said: 'Before answering them, we should talk among ourselves. We must keep our own way of organizing things.' When they had spoken among themselves, he said: 'Make it one *nigife*.' The *commandant* agreed.

Afterwards, the *commandant* said: '*Ce Monsieur, comment il s'appelle?*' He was told: 'Mammadu Denba So.' He looked through his papers, and said: 'Yes, your name is down here; you were the great spy. You're lucky; not long ago, if we had found you, we would have killed you. But I think spying is over now.'

Adrian Adams

From 1888, poll-tax was imposed upon Gajaaga as a whole. According to the colonial archives, it was difficult to collect. Efforts to adapt to the decline of trade by producing groundnuts and locally woven cotton for sale did not survive the construction of the railway, competition with cloth imported by French firms, and the abolition in 1905 of domestic slavery. Men began to seek work elsewhere: in the groundnut-growing areas of Western Senegal; on shipboard, first on the river, then from the First World War on ocean-going vessels; and in trade, especially in Equatorial Africa.

Once supplies were no longer needed for trading posts or army forts, the colonial authorities lost interest in existing River agriculture, except as a source of cash for poll-tax or food for the groundnut-growing areas. From the First World War on there was renewed interest in the prospects for irrigated farming. But in the end, the river chosen was the Niger; largely, it seems, because the presence in the Senegal River Valley of an active farming population was seen as an impediment to such schemes, as it was at the time of earlier attempts to create an agricultural colony.

Jaabe So

There were no more wars. But Sanba Haawa was still a fine horseman. One day, some Moors stole a hundred head of cattle from his *xaramoxo* in Jaagili. Sanba went after them on his own. He killed some of them; others fled. Three men had been left to guard the stolen cattle. He tied them up, and made them drive the cattle back to Jaagili. When they reached Jaagili, he took them to his *xaramoxo*, a Saaxo, and said: 'Here are your cattle; and three slaves, if you want them.'

That was when the people of Geyiga gave Sanba Haawa a large plot of land at Sanba-Salu. He had his slaves cut down all the trees near the ravine, where Seexu Mammadu Lamiina's men had waited for the troops from the fort in Bakel, so that Moorish robbers could not hide there.

In his old age, Sanba Haawa once pointed to an unmarked place in the burial-ground, saying: 'That's where I want my grave to be.' They asked him why. He said: 'Because those who died fighting for Seexu Mammadu Lamiina are lying there.'

On the day of my circumcision, Sanba Haawa set me upon his horse and rode with me through the town. He wept. He said: *'Duna ga maxa bono* . . . If only the world had not been spoilt . . .'.

When we cleared Seegankaani and Sanba-Salu, we found bullets of soft lead buried in the earth.

Amma So

Jaabe So

My father's first wife was Fenda Sawuneera from Soboku. She bore him Faatuma So. She also bore three sons, who died.

My father married Juumu So. Juumu So bore him four sons; not one of them survived to manhood. They all died at seventeen or eighteen years of age.

One of them went down to the river with some other boys; they were swimming, and diving after fish called *xooxa* that hide among the rocks. He put his hand in a crevice to catch a *xooxa*; a bracelet he was wearing caught in the crevice, and he couldn't pull his hand out. They looked for him; they saw his clothing, and other boys said he'd been with them in the water. They found him with his hand trapped in the rock, and brought him up, dead.

Another was fishing by the river, when a fish, an *anjoobe*, leapt up and bit him in the neck. His blood flowed inwards. All Kuŋani came down to the river in sorrow. Fode Mammadu, Seexu Jomo's son, came down from Xoje in a white robe, and took the boy in his arms. The robe was spoilt. The *moodi* wept. You would have thought Kuŋani had been destroyed that day.

Another had a headache, just for three days, then died. Another fell ill with guinea-worm. He also died.

My father was already an old man when he married my mother. Seexu Jomo told him: 'I've found a girl for you in Manayeli; I've asked them to give her to you.' At first my father refused.

My brother Siixu and I were the only two of his sons to survive to manhood. After Siixu, my mother bore a daughter, Buja. She was born with teeth, and kept hidden; then people said my mother should take her across the river the Jaajebinne, where she had a sister. She stayed in Jaajebinne until Buja was weaned, then left her there with her sister. I saw Buja several times. She was beautiful, with a forehead that shone like a mirror, and eyebrows that reached her temples. She died when she was eight.

I never really knew my father, as he died when I was only four or five years old. What I know of him I heard from Juumu and my mother, and from Sanba Haawa and Saada Sire Timmera.

Sanba Haawa was my father's friend. He was a great horseman, a great warrior. He and my father fought for Seexu Mammadu Lamiina at the battle of Sanba-Salu; Sanba Haawa himself told me about it. They fought at the battle of Bulebone.

My father was a great farmer. Of course he took part in the affairs of the town; he counted with the *debegume*. But he was a reserved man; he would speak only at important meetings. People would ask him to intervene if there was a problem they couldn't resolve.

My grandfather, Mammadu Denba, never farmed; he had herds of cattle, and every year he would go and collect his supply of grain from the Bacili in Tiyaabu. But my father, Amma So, was a great farmer. He provided for his household by farming alone.

He would farm three acres of a piece of land he had at Fanqanne, close to the town. At that time, there were seventeen acres there in all; he would lend the rest to other people. He farmed the same land every year.

A month before the rains, he would clear his field. That took him nearly a week. When the weeds had died, he would burn them. Then he would till the field with a short-handled hoe: not a large field, as he was on his own. From then on, he would go to the field after early morning prayers, returning only at sunset. His wives would take him food, each in turn. When the rains came, he would plant maize. When

By the River (c.1720–1933)

the maize was nearly ripe, he would plant *gajaba* sorghum in the same field.

Everyone admired my father's way of farming. When the ears of maize were still young, he would ask Seexu Jomo to lend him some *taalibo*, and they would weed the maize-field until it was perfectly clean. Then he would sow *gajaba* under the maize plants. When the maize was ripe and dry, he would harvest it and stack it on high ground, then weed the *gajaba* once more. All his life he farmed like that: maize and *gajaba*. There was a *kome* here, Musa Juma's father Kesba, who grew *fuñanŋe* all his life. My father gave him land at Fanqanne, and grew a small plot of *fuñanŋe* alongside him.

Sanba Haawa told me how at the end of the rainy season, when my father was planting maize and sweet potatoes and cowpeas in his *falo*, he would go to the field to chat with him, tethering his horse under a shelter my father had built by a *turo* tree. There would be a fine stand of *gajaba* then, as far as the eye could see.

After my father had harvested the *gajaba*, he would continue work on his *falo*, planting calabash and pumpkin lower down on the slope as the river receded. When he finished, there would be only two or three months left before the start of the next rainy season.

The maize he grew filled one whole granary, and the *gajaba* one and a half. It was more than his family could eat. He would keep one granary of *gajaba* until the hot dry season. Foode Mammadu of Xoje was his friend. He would tell him: 'I have some *gajaba*, you know.' If the *moodi* had money, he would send *taalibo* to thresh and winnow the *gajaba*, then measure it and pay for it in cash. If the *moodi* hadn't any money, my father would say: 'I'll lend you the grain; you have people to feed.' After the rains, when the *moodi* had threshed his harvest, he would send my father the grain he owed.

My father farmed like that throughout his life. That is how he bought clothing, and gold for our mothers; that is how he bought horses. All his life, he never even went as far as Tambacounda.

Other people tried to farm in his way. My father himself told Sanba: 'Plant your maize in such-and-such a place, and when it's nearly ripe, let me know.' When he came to look at Sanba's field, he said: 'No, Sanba, don't leave it like that; you must clear the weeds completely, so the ground under the maize is clean, before you plant the *gajaba*. If you plant the *gajaba* among all those weeds, it won't do well.' Sanba said: 'It's not possible to till the field with the maize so tall.' My father said: 'Just wait and see.' He planted the whole field with *gajaba*; they

> In the River area, the famine resulting from the failure of the 1913 rains has had even more serious effects than had been foreseen. People have had to pawn all their belongings (jewellery, herds, land) in order to procure basic necessities.
>
> (Archives of Senegal: Rapports politiques trimestriels, 2G 14/6)
>
> There has been a fine harvest in the River area. During 1916, the colony exported about 5,000 tons of sorghum as fodder for the French cavalry.
>
> (Ibid., 2G 36/3)
>
> It is precisely in the areas where irrigation is possible, that the land belongs to the natives. They will never consent to give it up.
>
> (Archives of Senegal, General Report on Agriculture for 1919)

harvested the maize, then weeded the field as he had told them. They harvested *gajaba*, although not as much as he did.

Madi Moodi Geyi tried it too. But no one could get such good yields as my father. This was because he would break the male flowers off the ripening maize; and also because he cleared the ground thoroughly under the maize, so that the young *gajaba* plants could breathe.

Sanba Haawa said that he tried it himself in the end, but he couldn't manage it. To grow a large field of maize, then clear it of all the late weeds—he couldn't do it. No one could.

They say that when my father had grown old, my mother kept her hoe hidden in his field; when it was her turn to take him food, she would spend the day helping him farm, because he was on his own. My father would tell her: 'Daado, God made me a farmer. If you help me, we'll harvest what we harvest. If you don't help me, we'll harvest what we harvest.' My mother would say: 'I've just come to take your mind off work for a bit, so you can lie down and rest.' She did that all the time, until my father died.

Adrian Adams

In Amma So's day, riverside land was owned much as it is today. At Fanqanne, the founders of the town now have about twenty-five acres of land, fifteen belonging to the Geyi household, and ten to the So. Twenty-one Tanjigoora households have fifty-eight acres, eight of which belong to Xoje. Six other *moodini* households have forty-two acres. Three *tunka-lenmu* households have eighteen acres; a leather-

By the River (c. 1720–1933)

workers' household, two acres. Five acres belong to a household of former slaves. So the *moodini* households, with the leatherworkers' household attached to them, have one hundred and two acres of Fanqanne; and the *tunka-lenmu* group of households forty-eight acres.

Much of Fanqanne has been lost to the river. About fifty years ago, Al-Haji Haamidu, Seexu Jomo's grandson, had his *taalibo* clear the land sloping down to the pond of Habalu; several households were given small plots there. At about the same time, people also began to clear land at the top of the inland slope, at Gunba.

Downstream from the town, at Sanba-Salu, the Geyi have twenty-five acres; five small plots, ranging in size from half-an-acre to one-and-a-half, belong to two Tanjigoora households, two other *moodini* households (including the household of Salle Dukure, who left Gori with Seexu Jomo), and a *tunka-lenmu* household. Sanba Haawa's household have the sixteen acres given him after he brought the stolen cattle back to Jaagili. In a second group of plots, Xoje has two-and-a-half acres, a leatherworkers' household two acres, and two other Tanjigoora households half-an-acre each.

The land further downstream was not part of that granted to Buleeli Denba So and Muulu Geyi, but was given later by the Bacili. The So household own the eight and a half acres the Bacili gave Mammadu Denba when he married Mariyam Siise; Muusa Kaba's household has eight acres, the *Bundunko* household seven acres, and two *tunka-lenmu* households nine acres. That makes eighty-four acres of riverside land downstream from Kuŋani: forty-three acres belonging to the *tunka-lenmu* group of households, and forty-one to the *moodini*.

Seegankaani was also not part of the land granted to Kuŋani. After Seega's death the So family did not use the land, and their claim was forgotten. In Amma So's day, the Bacili gave Seegankani to Waali Silaamaxa Sisoxo, a *tege* from Bakel.

These are rain-fed farmlands, occasionally flooded. There are also floodlands by the river: eleven acres of *falo* on the inner slope of the river bank: eleven acres in all, four acres belonging to the *tunka-lenmu*, seven to the *moodini*; the *kolangal*, land farmed when flooded by the creek that flows in the ravine between Sanba-Salu and Seegankani, two hundred and fifty acres at the utmost. It was in Mammadu Denba's day that the *kolangal* was first cleared and shared out; it was shared out once again in Jaabe So's youth. Its two hundred and fifty-odd acres are divided into sixty-eight plots, forty-five belonging to *moodini* households, five to *tunka-lenmu*, and seventeen to households which, with one exception, own no land at Fanqanne or Sanba-Salu: households descended from slaves.

Leaving the irregularly flooded *kolangal* out of account, Kuŋani has two hundred and forty-five acres of riverside farmland. Amma So farmed only at Fanqanne, and provided well for his household; but he was one of the few to farm only by the river. Kuŋani's main farmlands are inland in the *jeeri*, where long-established households are joined by those as yet off the map; those who may well be given land to farm at Fanqanne, but who do not own it.

Jaabe So

I also heard about my father from Saada Sire Timmera, who was then *chef de canton*. I knew that Saada Sire had been my father's friend; he knew my elder brother Siixu, but I'd never met him.

One year when I was a youth but not quite grown up, about sixteen years old, it happened that just before sundown on the third or fourth of the *Baano* feast-days, Mammadu Laaji Tanjigoora was brought a message to send on to Saada Sire. That was in colonial times; when the *debegume* received a message, the person whose turn it was was given the message to pass on.

Mammadu Laaji found us youths amusing ourselves. He said: 'Who will be so good as to take this message to the *chef de canton* in Yafera?' At first, no one spoke. His own son was there; he was about my age. He said once more: 'Who will take this message to Saada Sire? It's just come, and they say it mustn't spend the night here.' No one spoke.

He then said: 'Youths, if one of you should die, his corpse would feed a vulture. Anyone who has enough flesh for a vulture, is of a size to be a man. Who will take this message to Saada Sire in Yafera?' I said: 'I will.'

He said: 'Who's that? Is that Jaabe?' I said yes. As we left, he said: 'Take my horse.' I put a bridle on the horse, but I didn't saddle it.

I set out. When I came to Golomi, they were starting sundown prayers. When I came to Yafera, I tethered the horse to a wooden platform by the entrance of Saada Sire's house. I went in, holding my spear, and found him lying in a deck-chair. I went up to him and gave him the message, saying: 'They told me to give you this.' He said: 'Don't you know you aren't supposed to bring a spear in here?' I said: 'I didn't know; besides, if someone is sent through the bush at this time of night, he's bound to take along something made of iron.' He said:

'No spear is allowed in here.' I said: 'I heard you.' He then asked whose son I was in Kuŋani. I said: 'I'm Amma So's son.'

He said: 'Is Daado your mother?' I said yes. He was silent for a moment, then he said: 'Your father was a man. Sit down.'

I sat down. He called his wife, and said: 'Bring food for him; this is my son.' They brought *futo* and milk, and I ate a little. I was in a hurry to get back, not because I was afraid of travelling by night, but because there was drumming and dancing in Kuŋani that evening.

He said: 'Here is five hundred francs; give it to my wife.' I said: 'Who is your wife?' He said: 'Your mother. Your father was like an elder brother to me, or a father. When I was made *chef de canton* here, the Tunka was against me, Kuŋani was against me, Golomi was against me, Yafera was against me. I was brought in against their will. Then all Kuŋani held a meeting, to say that the town was their own; they would not obey the colonial power. Saada would not set foot in Kuŋani. Your father was at the meeting, just listening. When they'd finished, he said: "I'll not turn a *mange* out of my house, nor a Bacili. The Bacili gave us Kuŋani. If a Bacili or *mange* comes here, I can only bid him welcome."

The *debegume* said: "Fode Mammadu is here; Al-Haji Buna Jaaxo is here; Al-Haji Al-Xaali is here; I myself am here. The whole town has agreed." Your father said: "Whatever the whole town has agreed, has to do with the town. You, *debegume*, do not rank higher than me in this town; and my house is my own. If Saada comes here tomorrow, I will open the back gates of the town and let him in."'

He left the meeting. When Saada heard of this, he'd been in office for over a year. He sent a message to my father, saying: 'Tomorrow, I'll be coming to spend the day with you. Don't go to any trouble. If you like, we'll bring our own food.' My father replied that if that was their plan, they'd best stay away. If they were coming, they should just come. No one ever went hungry in his house.

Saada set a date for his visit. My father organized things well. He called his slaves and had them make a quantity of *fonde*. The sugar available at that time was called *beke*: whole sugar-loaves wrapped in blue paper. My father bought sugar; he had milk. They cooked the *fonde*. He went to welcome Saada and his people, and escorted them directly to his house, without going through the town. When they had eaten the *fonde*, he had an ox brought before them, saying: 'Saada, this is my gift to you and your people.'

The ox was slaughtered. In those days, there was no rice; the slaves who accompanied Saada, the other people who accompanied Saada all ate meat and *futo*. In the afternoon, people were given meat to take home.

Saada told me this, Saada Sire Timmera. Afterwards I asked Sanba Haawa about it, and he said that was exactly what happened. I said to Juumu, 'Aunt, so Saada used to come here, even though Kuŋani had said he mustn't come.' She said: 'Yes, there was trouble about that at the time. People spoke to your father, saying that the whole town had decided Saada wasn't to come here. He just said, "He's coming." Saada did come, and he kept on coming here until he and the people of the town were reconciled.'

Our connection with Mahamme Kamara's household dates from my father's time. Mahamme's grandmother remarried when his fathers were little. She married Muusa Lenme's grandfather, and went to live with him in Tanjankunda. Her sons grew up there, with her husband's sons, until they were youths.

One day when they were out in the fields, they had a fight; the husband's sons injured the wife's sons. A meeting was held at the Tanjankunda mosque. At that time, Madi Moodi was *debegume*; he would do nothing unless my father was present. The boys were reconciled. But the year after, they fought again in the bush; Madi Siise, Mahamme's father's elder brother, was injured. People brought him back to town, and a meeting was called at the Tanjankunda mosque to deal with the matter.

Muusa Sunbulu was one of the elders of Tanjankunda. He said: 'Here come Amma So and Sanba Haawa; let's wait for them.'

Amma So said: 'What's happened?' He saw Madi Siise sitting there bleeding, and said to the *debegume*: 'Namesake! Ammadi!' The *debegume* said: 'Yes?' My father said: 'If you're not careful, I'll have you put in prison.' The *debegume* said: 'Why?' My father said: 'These people aren't related; let's separate them.' The *debegume* said: 'Where shall we put them?' My father said: 'In the town square.'

They built them rooms next to where Condi's shop is now, and they went to live there. They weren't in the slaves' neighbourhood any more. After a while, Amma So said: 'Let them join us, and take part in our meetings. A person who has no one of his own, belongs with the *debegume*.' They came to live across from our house.

A woman slave named Ñaxana, belonging to Bubu Dabo of Baalu, ran away with her baby and sought refuge in my father's house. My father wouldn't send her back against her will; when Bubu Dabo heard that, he said: 'Let her stay.' She became part of our family. My father gave her land to farm; she was friendly with Fode Mammadu, and he would send his *taalibo* to help her.

Her daughter Kebe Ñaxana married Mahamme Kamara's father. In addition to the land her mother farmed, she also farmed a plot further upstream, belonging to the Geyi. After her husband's death, the *debegume* wanted to take back that land; she came to see my father, and he said she should be allowed to keep it, as she had no children of her own. That is the plot of land Mahamme Kamara owns now.

Waali Silaamaxa Sisoxo was my grandfather. My mother's paternal grandmother came from their house; he counted among my mother's fathers. He was also my wife Maamu's real grandfather, her father's father.

Waali Silaamaxa came from Xaayi-Xaaso to settle in Bakel. He forged hoes for the Bacili's slaves; they asked him to make bullets for them as well. He made sacks full of bullets. The Bacili asked him if he wanted slaves as payment. He said no; he wanted land. That was when they gave him Seegankani. At first he farmed the land; he had many slaves. But then he went to work on the Koulikoro railway line. He became a skilled worker, a mechanic. He gave up farming and took his family to live in Kayes, Koulikoro, Bamako, Dakar, Thiès. The land lay fallow for fifty years.

When my father died, he left two full granaries. At the funeral, Foode Mammadu said the grain should be sold at the best possible price, to help care for the orphans. We were orphans because our father had died. He entrusted us to our *ka-tugune*, the younger branch of our family, Juumu's father's household.

Juumu never remarried. After my father's death, Sanba Siise from Bakel wanted to marry my mother. But she said she wouldn't go to Bakel; she would stay in our house, and keep a fire lit there for her children. Foode Mammadu said to her: 'You're still young.' She said: 'Don't worry, I'm not going to play around; I'm not going to bear any bastards.'

After a while, Foode Mammadu said Daado should marry Bakari So, Moodi So's son. She didn't want to. There were two people she heeded in Kuŋani, Sanba Haawa and Foode Mammadu. Sanba Haawa said to the *moodi*: 'Since Daado refuses, why not leave her alone?' But the *moodi* said: 'No, we mustn't leave her alone. Through her, two households will be saved.' He knew what he was saying.

She finally agreed to marry Bakari So, but she refused to leave my father's house; Bakari So would spend the night with her there. The year she married Bakari So, he went to Dakar, as did his brother Malangal. Bakari So caught the plague in Dakar; he went to Tambacounda, and died there. Malangal So hid on board a ship in order to go to France. He hid in the coal compartment in the hold, along with NpaBakari Jabira. There was no air in there. The ship spent the night in port; they battered the hold with their fists, hoping someone would hear them. At last people came by and said: 'There's someone knocking.' They went to fetch a key. When they opened the door, they found Malangal dead. They took NpaBakari to hospital. At that time, Daado was pregnant with my half-brother Waagi. When Waagi grew up, he built his father's house anew.

At first, my mother gave me and Siixu to Juumu, because Juumu's children had all died. Later on, she gave me to her elder brother in Manayeli.

After our father's death, our mother fed and clothed us. She farmed, and she made pots. She would take shea-nuts, roast them and grind them to a powder, then work them to a paste and add perfume. She would make that paste into small beads, pierce them and string them into necklaces; the fragrance lasted forever. She made water-jars and bowls of potter's clay. And she grew maize and *gajaba* sorghum in my father's field at Fanqanne.

She was living in our father's house. Every year when the rainy season came, she would tell our other household: 'I will provide the grain and groundnuts for the next three months.' She would measure it out and give it to them. Daado did all she could. And we did not let her down. From when we were young, we did well whatever we had to do, farming, fishing, or whatever. No one could ever find fault with us because we were raised by a woman.

My mother died last year; she was over a hundred and ten years old. I give thanks to God.

Yugo-xase Gangaji

Jaabe So

My mother's side of the family: that's a large subject! My mother's father was Yaaya Koli Gangaji. His family was from Jaagili; they were Gidimaxa people. All the Jabira called them *Lebekebenke*. All *Lebekebenke* are of the seed of Yaaya Koli, of Koli Maxan. They were the great *ñaxamaala* of Jaagili, the great casters of spells. They came to Jaagili—but I won't go into how Jaagili was founded. That's another story, that follows my father's path, the Bacili path. My mother's is the path of the *Lebekebenke*.

In those days, the only horsemen and warriors of Gajaaga were the *tunka-lenmu*: the Bacili, their *mangu*, their *sakko*. All along the river, none but they were called horsemen or warriors, or masters of the land. Anything else is a lie.

After a battle at Manu-Jeeri, the people of Manayeli came to Jaagili looking for someone to cast bullets for them. They said to the people of Jaagili: 'Let us have Yaaya Koli to make bullets for us.' The people of Jaagili said: 'We'll let you have him, on condition that you treat him well. If you don't treat him well, we'll go and fetch him back again, because of the bond between *Lebekebenke* and *mangu*.' Yaaya Koli went to settle in Manayeli. In those days, he made countless bullets for the Bacili wars. When the *tubab* came, they fought from Hayire to Manayeli. In Manayeli, they fought but did not win; they fought in Tiyaabu but did not win, thanks to the people of Manayeli who went to help Tiyaabu. Out of all Gajaaga, those are the only towns the *tubab* did not defeat.

Yaaya Koli sought the hand of Liŋu Saafa, Saafan Sallo Faadiga's daughter. Saafan Sallo's seed are from Yelingara. They too were *tego* and casters of spells. They could present an ox to honour a guest, slaughter it, eat the meat, then set aside the head; the head would become a whole ox. All Gajaaga knows that. They were *tego*, but magicians as well.

Yaaya Koli married Liŋu; Liŋu bore him our mothers. She had eight daughters. They all married and bore children; the shortest-lived of them reached seventy years of age. Gidimaxa, Gajaaga, Fuuta, Mali: no one has more family than we do. All of Liŋu's daughters had seven or eight children, more often ten or fifteen.

That was how my grandfather went to live in Manayeli. But he still had close ties to Jaagili. When, in the end, his son Yugo-xase went back there to live, all the slaves and *mangu* of Jaagili beat the drums for him; they built him a house and tilled his field.

For a long time, my mother's brother Yugo-xase Liŋu did not have a son. So my mother gave me to him. I became his eldest son. When I was working abroad, I sent him money and bought him horses, all his life.

My uncle's first wife was Tako Booyi. She was my grandfather Saafa's daughter. My uncle went to France, and left me and Tako Booyi at home. Every morning, Tako Booyi would take his horse down to the river and wash it. Every rainy season, Tako Booyi and I would clear a field and sow it with sorghum. Tako Booyi stood in my uncle's stead just as a son would have done.

I was young then, but Tako Booyi and I did everything together: till, sow, harvest, plant the *falo*, and fetch dry grass from the bush for the horse. Tako Booyi would take a rake to gather the dried grass and tie it into sheaves. She died in 1938. My uncle was in France then; he came home, and did not leave again. That was when he gave up working on board ships.

Sallo gave him Tako Booyi's younger sister, same father, same mother: Maama Booyi. She bore him two daughters, Xunba Maaxo and Faatuma Maaxo. Tako Booyi never had any children. There were few women like her to look at.

Maama Booyi also died. When my grandfather heard this, he gave my uncle another of his daughters. She bore him a son, Jaabe. But my uncle was not very fond of her, and they were divorced.

I was brought up by my mother's brothers: among them was Sanba Malowu. Everyone from Keniyu to Gande knew Sanba Malowu, every *Tunka* from Abdu Salaam to Konko Goola knew Sanba Malowu. He never lied; he showed neither fear nor favour. He was a *sakke*; the Bacili feared him because of the bond between them. The Manayeli *mangu* would ask his advice when they were in difficulties. He was a wise man.

If a woman left her husband's house, Sanba Malowu would not go after the woman to persuade her to return; he would send for her, and ask her what the trouble was between her and her husband. Once he had reconciled them in his house, he would send for the woman's father. He himself would not go out. There were three date palms in his back courtyard, with four mats spread underneath. From *laxa-*

By the River (c.1720–1933)

asara onwards, there would be people sitting there; when one left, another would arrive. He spent his whole life like that, setting things right.

There was more democracy in colonial times than we have under Independence. In colonial times, even if someone killed someone, the administration would say: 'Settle the matter among yourselves; if you can't, come to us.' With today's rulers, even if you insult someone's mother, the administration will make it its business. *Ken waxati, demokaratiikin wa du yi*. In those days, democracy was natural.

Sooner or later, a day will come when no one will listen to those who've been to school, when they'll turn back to their forebears. We have our own *fosuwilisaano*, our own *taarixu*. If Sanba Malowu were alive today, he could prove that we and the Moors have never shared a common border. He could recall the time of the first Moor who came down to the river from the Sahel. His name was Moxtaar uld Seyin; he asked the rulers of the country for permission to graze his herds. He had large herds, and many Hartaan, black Moorish slaves. *Koliyaajo* would travel from Fuuta to the far reaches of Gidimaxa, to hire them as mercenaries paid with silver and gold. The Bacili too would hire them to fight for them.

In those days Testaayi did not exist; people from Manayeli would go and farm as far as Wuluraani, twelve miles north of the river. We had nothing to do with the Moors then. Now liars say the river is the border between us. A day will come when the River people no longer heed the educated people of today, but follow our own buried *taarixu*.

I was born in Kuŋani, but I have family in Golomi, Jaagili, Bakel, Tiyaabu, Manayeli, Yelingara, Jawara. My mother's mother was from Yelingara; my father's mother was from Bakel. When I was a child, after my father's death, Konko Goola was Tunka. He would send a message to my mother, asking her to let us come and talk with him. *N fa* Konko would tell us: 'You are the masters of Gajaaga, *sakko* and *mangu*. Gajaaga belongs to you. Your fathers and grandfathers were organizers for the Bacili. You are the Bacili's companions, at home and abroad.' I would spend two months with the Tunka in Tiyaabu, then return to my mother.

Then I went to live with my uncle Yugo-xase in Manayeli. My uncle's close friend was Usuman Booyi in Mudeeri. Every year he would go and spend three months in Mudeeri, working as a *tege*. When the people of Mudeeri harvested their *kolangal*, they would collect grain and thresh it, then load many, many donkeys. The *komo* would take Yugo-xase's grain to Manayeli. The day they arrived, he would slaughter an ox in their honour.

My uncle Yugo-xase grew three crops a year. During the rains, we grew *ñobugu* sorghum at Sanbanpare, just outside Manayeli. Every morning I went to that field, to stop birds and ground squirrels from digging up the seed. Then we weeded it. At harvest time we stacked the grain, and gathered the cowpeas we had planted along with the *ñobugu*; then we gathered in the grain.

My uncle also grew maize on a large *falo* he had on the north bank of the river, at a place called Lesaanu, just across from the pillar at Sibiko, between Tiyaabu and Manayeli. I would go and keep watch over that maize, from the time when it began to ear until harvest time. In those days monkeys would do a lot of damage in the fields. My uncle's double-barrelled gun was too heavy for me to lift and shoot. I would put two forked branches upright in the earth, with a third branch laid across them. When a monkey came near—sometimes it seemed about to attack me, and I was frightened—I would rest the gun on this stand, and fire; the monkey would flee.

Then we would sow the *kolangal* with *sanme* sorghum. In those days the *kolangal* was regularly flooded. That field too had to be guarded against monkeys and birds. There were three kinds of farming during the year: the rain-fed field, and the two kinds of floodlands, *falo* and *kolangal*. There were four of us, with my uncle's brothers Saliki Gaajigo, Denba Labbo, and Mammadu Fenda. I was a boy; Saliki and Denba Labbo were youths. My uncle was then about fifty years old. He harvested so much grain that the entire household—two men, seven women, two youths, two boys, three small children—would not finish it in a year. We ate, the horse ate, the sheep ate, and still one year's grain would not be finished by the next year's harvest time.

I spent all my time at Lesaanu guarding the maize, with Moyi and Sanba Komo's son Madi Moodi. We ate together, and rested together under the same shelter. In those days there was a pond nearby; in the heat of the day, when the parrots slept and the monkeys stayed away, we would take our nets and catch fish for our lunch. Madi Moodi would cook the fish, because his field lay between ours, and monkeys didn't come there. After we'd eaten, they would send me to fetch water. When the level of the river fell, it was a long way to go; I told

By the River (c. 1720–1933)

them that if I stayed away from my field any longer, the monkeys would come back. So they started fetching water themselves.

They weren't pleased. They couldn't refuse to eat with me, because the food from our house was good; every morning we would eat what our household had cooked the evening before. But one afternoon, instead of calling me to cross the river with them in the canoe we used, they left me behind. I had no way of crossing the river; so I went to spend the night at the shelter Sunsu Boojo of Tiyaabu had built for the farming season. There was a boy of my age there, Denba Sunsu; my elder sister Xunba Maaxo was there, and Budulu Jaare also spent the night.

On the second day, they left me behind once more; and again on the third day. I decided: 'I'm not going to spend the night here. They'll have told my uncle, "Jaabe went to spend the night at Sunsu's."' At that time the maize was ripe, and I thought I'd be able to ford the river.

I had a dog with me, called *Maxa-mungu-tege*, Blacksmith-don't-forget. I took off my loincloth and put it in a bowl, on top of two roasted ears of maize.

I began to cross. The water reached my neck; then I swallowed water. I held onto the bowl with one hand and began to swim using my other arm; when that arm tired, I switched to the other.

Then I saw a crocodile coming towards me, cleaving the water. When it drew near me, it dived. The dog knew it was there, and tried to come closer to me. But the crocodile caught the dog. When it turned, its tail struck me and I went under. I lost the bowl. The crocodile went off with the dog. I swam and swam.

I was near this side of the river, when I saw another crocodile coming at me from a distance, cleaving the water swiftly. There was a man from Tiyaabu standing on the bank, and he shouted at me: 'Hurry! Hurry!' I was swimming away; I didn't know I had reached the bank until my foot struck solid ground. I stood up; the water was waist-deep. The crocodile was near me by then. The man from Tiyaabu came and caught hold of me.

We'd nearly reached Tiyaabu, when I said: 'I'm bare-arse.' He said: 'You could have been dead by now, and you're worried about being bare-arse?' So I covered myself with my hand. When we came to Kunda Sakke's house—it was the year Kunda got married—he said: 'Here is your son. God has preserved him. Give him something to cover himself, he's crying because he's got nothing on.' They wrapped a cloth round me, and the man told his story. The house was full of people.

A fisherman who had been near the pillar had gone to Manayeli that evening. He told people there that a young crocodile had gone after a boy; his dog was taken, but the boy escaped.

My uncle said: 'That's Jaabe.' He and Siyaaxa Cici set out on horseback. It was night by then, and you could hear their horses' hooves as they neared Tiyaabu. My uncle came to the house where I was, and set me behind him on his horse; we went to spend the night in Manayeli.

He told off the other boys, saying the canoe was for all of us, and they shouldn't have left me behind. The Manayeli *debegume* said to them: 'You shouldn't have done that. You did it out of the unkindness of your hearts.' My uncle took some charms sewed up in leather, and tied them round my waist. He said: 'Next time, if you like, don't bother with a canoe either.' He was very angry.

One year I kept watch over the *kolangal* with Saalun Xulle, a young *kome* from Manayeli; along with Sanba Jara, and Maama Jeegi who later married the eldest son of the head of the Manayeli Koranic school. They were all my own age. Just below us was the pond called Maanu-Jeeri.

One day we were eating our morning meal. Saalun's father was in the habit of leaving his gun there, in order to shoot at monkeys that were spoiling the crop. When we had finished eating, Saalun got up and took the gun. He said: 'Look at me, Sanba!' Sanba said: 'You can't shoot; why don't you put the gun down?' Saalun said: 'Who, me? I'll just . . .'

He shot Sanba like a warthog, full in the chest. Sanba ran and fell into the waters of Maanu. We were only children; we started crying.

All the people came from the nearby fields, and dragged him out of the water; they sent word of what had happened to Manayeli. At that time, Konko Goola was Tunka in Tiyaabu. Saalun fled and sought refuge in Tiyaabu, because that is the custom between the *mangu* of Manayeli and the people of Tiyaabu. If you kill someone in Manayeli and seek refuge in Tiyaabu, you will be safe. If you kill someone in Tiyaabu, and seek refuge with the *mangu* in Manayeli, you will be safe.

The Bacili and the Manayeli *mangu* met. They kept things quiet, no one said a word, and Saalun didn't go to prison. He continued to live in Manayeli until he went to work in France, then married and had children. I witnessed that too in my lifetime.

2

Inland
(*c*.1720–1938)

Adrian Adams

Riverside land is farmed year after year. In a good year, there will be a crop of rain-fed maize on higher land, then a floodland crop from *falo* and *kolangal*. If the flood destroys the rain-fed crop, the same land will yield a flood-recession crop in its stead. But Kuŋani's main farmlands are elsewhere.

In a good year, the household to which young Jaabe So returned in 1934 might harvest forty *kande* of maize, about one and a half tons, and eighteen to twenty donkey-loads of *gajaba* sorghum, about four tons, from the land it farmed at Fanqanne. The So household, two men, three youths, four women, and two girls, needed three *muudu* of grain a day, under four tons a year, for its subsistence. It is one of the few households that could, in a good year, subsist on riverside farming alone: relatively small, with generous holdings of land. Yet they also farmed elsewhere.

The household to which young Issa Budu belonged, the household then headed by Sanba Koyita Tanjigoora, descended from one of the earliest Tanjigoora settlers, might harvest at most fifteen *kande* of maize, half a ton, from its riverside fields. They needed seven *muudu* of grain a day, nine tons a year, for their subsistence. They had to farm elsewhere.

There were many households less well endowed than theirs with riverside land. There were other households which owned no riverside land; they came into existence at a time when that land had already been shared out. And there were people living in households of the town, who did not farm the household fields.

Two hundred and twenty-five acres of riverside land will produce at most ninety tons of rain-fed and flood-recession sorghum and maize; about a third of the town's present needs. Even granting that Fanqanne was broader sixty years ago, the town as a whole, and most of its households, cannot live on the riverbank alone. Kuŋani's main farmlands are inland in the *jeeri*, where long-established households are joined by those as yet off the map: *komo*, descendants of slaves; *taalibo*, Koranic students; and women.

All these people farm land that lies two, four, even six or ten miles inland, south of the river, among low rocky hills. With a few exceptions—the land at Garsingide given to Buna Fasunte, and the land at Suxangide, even nearer the town, owned by the Geyi and So and shared out among other households—these are not permanent fields, the property of a given household. The land belongs to Kuŋani as a whole. Within it, groups of households clear fields in a given place, farm there for several years, then move on.

These fields leave even less of a trace than the tenuous boundary stumps and stones of Fanqanne. Their size depends on the strength and needs of those who clear them, and once abandoned they are soon overgrown. But they can be mapped from living memory. The palimpsest of riverside fields charts Kuŋani's origins. But the *jeeri* with its outcrops of rock, white-thorned trees and blond grasses, featureless and shifting like the sea, is the broader base from which people from Kuŋani ventured further afield: to the coast, to lands further south, to Europe.

Household Fields

Isa Budu

Garsingide, jeerin ya ni; Garsingide is in the *jeeri*, but it's ours. The people of Geyga gave it to Buna Fasunte. If someone wants to farm there, let him come; it's *jeeri*. If he likes, he can farm the land we're not using. But no one can farm there without our consent. In all the *jeeri*, we alone have land on those terms. *Alla siren ya ni.* God is good.

Al-Haji Ma Jaaxon Ba

Ke ñiiñe su, Fanqanne, Sanba-Salu, the *jeeri*: all that land belongs to God, like the sky. By the river, it's your own land, your father's, your grandfather's. The *debegume* gave land according to people's needs. Someone can ask you for land to farm, but it doesn't become his; the land given to your grandfather remains your own. But in the *jeeri*, the land doesn't belong to anyone's father or grandfather; you farm it for two or three years, then you leave, and someone else can farm that land after you.

Al-Haji Mpali Tanjigoora

Ganni, i wa soxono, i yi fo kitana soxoye. In the old days, people farmed and they made something out of it. The river flooded, and with the flood, fish went inland. That's what makes fish plentiful; when fish can go into the bush to spawn. After two or three months, the waters go down again; the fish go too, large and small. Some are caught; the rest return to the river. Fish are plentiful then.

In the old days, people didn't farm by the river. They farmed inland, in the *jeeri*, because of the floods. Our households were called *Bundunko*, 'the people of Bundu', because they used to farm at Lonboli. They didn't farm by the river. When the rains came, the whole household would go to Lonboli. They went there to grow maize, and *ñobugu* sorghum; their wives grew groundnuts. They also grew *fuñanŋe*. At the end of the rainy season, they would bring back with them what they needed to live on during the dry season; they left the rest in their granaries at Lonboli. The following year they would return to Lonboli. That was the way our fathers farmed; we ourselves were born to it.

The father we knew was the eldest, whose name was Ba Bintu. At that time our other fathers were dispersed abroad. We farmed one year with him at Lonboli, growing maize, *ñobugu*, *fuñanŋe*. When we returned here, the waters had gone down. Where *gajaba* grew best, we planted *gajaba*; where maize grew best, we planted maize. In the old days, people lived on their strength, on grain and fish.

In the *jeeri*, the first thing was to clear the field. The next thing, when it rained, was to sow it. When you finished that, you would come and till a plot by the river for growing maize. That was just so the children would have fresh ears of maize to eat; because we knew that maize-field would be flooded. It never failed; year after year, the river flooded. Then we'd go back to the *jeeri*.

Our *jeeri* farming wasn't a matter of going out in the morning and spending half a day in the fields! No. You pounded your grain, you prepared *futo* and *ñecce*, groundnuts, dried fish, and took it all with you to the *jeeri*, where you spent the night. One person would be sent in the morning to go and fetch the midday meal. They would prepare *futo* during the morning for him to take back.

At night you'd eat *futo fillante*, 'twice-cooked *futo*'. In the morning you'd eat *sonbi*, porridge. You'd carry on farming until the midday meal arrived. It was always the same thing: *futo dere*, *futo* with leaf sauce. That was what you farmed on. Ah! that *futo dere* is really healthy food.

After the year we farmed with Ba Bintu, I really started farming with our *kome* Baare Soxona. We farmed at Fendagesse; we were there for four years. Then we farmed at a place called Jamyeli; it's at Guñan Yaamadu, but its own name is Jamyeli. It's all one stretch of land, but each part has its own name. We farmed there for six years. All that time, we were with Baare Soxona. As I said, we spent the night in the *jeeri*. We didn't go out in the morning and come home at night.

When we left Guñan Yaamadu, I went abroad. When I came back, we went back out there, to Manga-Gufan-Xanne; that is also part of Guñan Yaamadu, but with its own name. We spent two years there, then the land went bad; *worowolle* started growing there. So we left.

I went abroad again, for three years. When I returned, I found that my father MaJaaxo had returned to Fendagesse. But we only farmed there for one year. The land was no good. I told my father we should leave it.

In 1958, we came to Lugere. We farmed there for four years. All this time, we grew the same variety of sorghum: *ñobugu*. Then I travelled abroad once more.

When I came back, we farmed at Janjume for seven years or so. My *jeeri* farming ended at Janjume. After that, I travelled again; after my return, I started farming at Sanba-Salu. Since then, I've not farmed in the *jeeri*.

Jaabe So

Siino fano, for the first few years after I returned to Kuŋani from Manayeli, my elder brother Siixu was still here. We grew maize at Fanqanne; at that time, the field my father used to farm was much wider than it is today. And we grew sorghum in the *jeeri*. When we farmed in the *jeeri* during the rainy season, in the early morning each of us went to work in his own small field. After breakfast, we would all work together in the *te xoore*, the family field. In the afternoons, when we'd finished working in the family field, we'd go to work in the field of the eldest man, Ancuman. After a while, Siixu and I would go and work in Siixu's field; after that, I'd work in my own field. Then Siixu would call me, and we'd wash and rest.

We guarded the maize at Fanqanne until it was ready to harvest. After it was harvested, we went to guard the *jeeri* fields at Guñan Yaamadu. In those days it was so pleasant out in the bush, that you didn't want to go back to town.

Inland (c.1720–1938)

In the *jeeri*, the strong trees were *kiide*, *ŋoñe*, *deye*; with *fa* and *sexenne* in rocky places. The *kolangal*'s strong trees were *jebe*, *xiile* and *fa*. Bushes were *xaame* and *sanbe*. The bush was thick with trees, each in its place. You couldn't climb hills for the trees.

There were guinea-fowl in the bush, wild ducks, wildcats and porcupines, to be killed and eaten. The bush was full of game. Nothing was lacking there. There were *siine* and *nese* in the bush, if you had a gun. If you had a machete, there were wildcats in abundance. If you had a dog, there were guinea-fowl.

While guarding fields at Fanqanne, at midday you could go fishing. In just five minutes with a net, you could catch any number of fish: *balde*, *sakinbaaya*, *gurlo*, *tallaqe*. You cooked them and ate them with *futo* or *ñecce*. In those days, those guarding fields in the bush needed nothing from town.

The harvest from the *te xoore* was for feeding the whole household. Siixu would store the grain from his own field in Juumu's granary. Juumu would tell me to put my own harvest in my mother's granary. My mother was a potter, and at harvest-time she would fill a small granary by trading water-jars for grain and groundnuts. She also farmed herself: she grew maize at Fanqanne, and a small plot of groundnuts at Selincina, a bit of land we had at the edge of town, so that her children would not have to ask other people for a taste of the new season's crop.

In 1936 there was a great flood that drowned the fields of Fanqanne and filled the whole *kolangal*. When the waters went down, Fanqanne was sown with maize and *gajaba* and *feela* sorghum, and the *kolangal* with *sanme* and maize. The Fanqanne ponds, Munune and Habalu, were full of fish, as were the streams. Even a boy could bring back enough fish for his family's meal. It was at the time of the 1936 flood that I first went away, to Dakar.

During the 1937 rainy season, I went to farm in the *jeeri*. Bullayi Ñuma was there, with Sanba Kamara, Nuru Tanjigoora, Mpalamaane Tanjigoora, Sanba Ñama Jimmera, Isa Buna, Denba Xulle; there were many of us farming in the *jeeri* there, at Guñan Yaamadu. From here to Guñan Yaamadu is about six miles, further than from here to Bakel. If you set out from here early in the morning, you'd arrive there at about ten-thirty.

Siixu was gone, and I was on my own that year. I cleared a large field. Afterwards, when Salun Geyi, the *debegume*, arrived from France, he came to see me and said: 'It seems you've cleared a field that's far too big for you to farm. Why don't we share it?'

I said no. I was deeply offended that people were saying I couldn't farm all the land I'd cleared. I had an idea. I went to see the blacksmith, and he made me an iron-tipped digging stick.

I went to sow my field. The ground was very hard and dry; I made holes in the ground with my digging-stick, and put grains of sorghum in the holes. I spent all day there; in the evening, I went to spend the night in Jaabalu, a Fula village not far from there.

I went out to the *jeeri* on a Tuesday, and spent the rest of the week there. I went back there the following Tuesday, with my digging-stick; I dug lots of holes, and when I'd finished, I put grains of sorghum in the holes. I then gathered leaves from a tree, and scattered them all over the field, to stop animals from taking the sorghum. I worked until I'd sowed the whole field.

After I'd finished, there was a whole month without rain. When the rains came, everyone went to the *jeeri*, to build a shelter for spending the night. We built the shelter, and everyone began sowing. I came back to town and went to till my maize-field at Fanqanne. When I'd finished tilling, I sowed the maize. Right. I went to the *jeeri* to see if my sorghum had come up. I found it had all come up, not a single sprout missing. The others were busy sowing.

No proper Soninke farmer will set foot in someone else's field once it's been sown. So those who were farming out there with me, the older men like Bullayi Ñunma, sent people to tell me off, saying I ought to come and sow my field. I went out to the *jeeri* on the day the whole lot of them had finished sowing and were ready to come back to town. Not to sow my field; just to have a look. Bullayi Ñunma gave me an earful: 'When everyone was out here, you stayed away. You're too stubborn. Where are you going to spend the night, all on your own? Jaabalu?'

I didn't say anything. I went to look at my field. It was fine. It had rained once; then there was a fortnight without rain. What they'd sown hadn't come up properly. The first rain hadn't wet the soil enough; they sowed all the same, between the first and second rains. The second rain was very heavy, and kept their sorghum from sprouting properly. When the second rain fell, my sorghum was already a hand's breadth high.

They went to sow again; and I went to weed my field. They thought I was sowing, but I was weeding. I was young.

Right. We carried on like that. Before they'd sown half their field, I'd finished my first weeding. I came back to town to weed my maize-field at Fanqanne. Then I went back to the *jeeri* to continue weeding there.

Before they'd finished weeding their fields for the first time, I'd done a second weeding. One day, some people said: 'So's field over there is looking very green!' The others said: 'It's the weeds!' One evening Al-Haji Ma-Tanbadu, went to have a look.

We spent the night out there. It was organized that way; everyone stayed out there during the farming season. Tuesdays we would go to the *jeeri*; we would come back to town on Sunday evening, and rest on Monday. We all spent the night out there, each with his mosquito-net. We'd build a shelter for when it rained. We did the cooking, just us men. At five in the morning, prayer-time, people would cook *sonbi*; we'd have some *sonbi*, then each of us would go off to his field, before sunrise. After the sun had risen, we'd gather to eat again, then go back to work.

At midday, we'd prepare a meal. There was dried fish, milk, *ñecce* and *futo*; *futo* for eating with milk, and ordinary *futo* for eating with meat or fish. We ate, then said prayers at two o'clock. At two-thirty we'd start work again, until seven in the evening. At seven people would gather: to wash, light fires, and prepare the evening meal. We kept on working like that until we'd finished farming.

We harvested the *jeeri* fields and stacked the grain. We brought the grain from the *jeeri* into town. In those days, whoever farmed the *jeeri* would harvest enough to feed his family. In 1936 I had eleven donkey-loads. Heads of households had twenty-five, thirty, or forty donkey-loads. People like Bullayi Ñunma had forty-five donkey-loads. In 1937 I was on my own, but I harvested two tons two hundred kilos of sorghum. I also harvested twenty-five *kande* of maize, about five hundred kilos. That was my last farming season.

When the 1937 rainy season was over, I went back to Dakar. In 1938 I worked for the Shell Company as a labourer, filling petrol drums. In 1939 I joined the Navy.

Adrian Adams

The last year Jaabe So farmed in the *jeeri* was 1937. In 1938, 'the year before the call-up', he was working as a labourer in Dakar; but there were many others left in Kuŋani to farm the *jeeri*.

That year every household in Kuŋani farmed a *jeeri* field: eighty-three households, about two hundred and fifty men, not counting youths.

Two Tanjigoora households, seven men and some youths, farmed at Gidilenmude, near Garsingide, about two miles inland; among them the household headed by Sanba Koyita.

Seventeen households, fifty men, farmed at Papata, three miles inland: six Tanjigoora households, including one belonging to Xoje, headed by Fode Mammadu, and its *almuudu*, younger Koranic pupils; two Kebe households, the Jimmera household of *garanko*, and eight households of *komo*. Fode Mammadu's *taalibo* also farmed a large field there.

Nine households, thirty-four men, farmed at Lugere, over four miles inland: five Tanjigoora households, including one belonging to Xoje; three other *moodi* households, one Daraame, two Tanbadu, and one *kome* household.

Three households, eleven men, farmed at Gece, six miles inland: the Gunjam household, and two *komo* households.

Twenty-four households, seventy-five men, farmed at Guñan Yaamadu, six miles inland: the Geyi household; the So household; four Tanjigoora households; three other *moodi* households; a Jabira household, two *garanko* households, and twelve households of *komo*.

Twenty-eight households, seventy-two men, farmed at Fendagesse, not far from Guñan Yaamadu: ten Tanjigoora households, including two of those descended from Al-Xaali Madi Bidan, and the *Bundunko* household headed by Ba Bintu Tanjigoora; two other *moodi* households, a Jabira household and thirteen households of *komo*.

There was a good harvest. Supposing the area farmed by each man to change little from year to year—indeed, it must have been greater at a time when the rainfall made good harvests likely—the men of Kuŋani will have farmed that year about 900 acres of *jeeri* land alone, and harvested over 300 tons of grain; enough for each to feed his household, and enough for the whole town's subsistence even fifty years later. Men travelled abroad, but they left fathers, brothers, and sons to farm.

Al-Haji Mpaali Tanjigoora

I su da julaaxun dabari; our grandfathers, our fathers, all worked as traders, but they all ended up farming. The only reason we left here was to seek our fortunes; to earn money, but always, in the end, to come home again. And whoever returns home must farm. It is farming that makes life good here.

Komo

Adrian Adams

Among those of Sanba Koyita Tanjigoora's household farming at Gidilenmude in 1938, were Isa Budu's father Abdu Koyita; two of his sons, an elder and a younger brother of Isa Budu; ten or so youths, and two *komo*. Certain other households farmed with former slaves that year: the Daraame, the Gaku, Fode Sire Tanjigoora's household, the *Bundunko* and a Tanbadu household. But these *komo* had their own plots in the *kolangal*, their own households in the town. Some had the use of plots of riverside land. Their sons did not work with their former masters, but farmed their own *jeeri* fields.

Thirty-odd households descended from slaves were established as autonomous by then. Three-quarters of these were former slaves of *moodi* households, more than half of them Tanjigoora. Perhaps a third of them have ancestors brought to Kuŋani over two generations ago; the others were brought in Samori's day, at the end of the nineteenth century.

Isa Budu

Before the *tubab* came, my grandfather, his brothers and his sons used to go and buy slaves. They would bring them back here, and send them to farm the land; they themselves farmed with them. They would go and bring back three slaves, then go again and bring five. After a year, some ran away and were caught; some escaped. Those who remained farmed with my grandfather, his brother, and his sons.

Buna Fasunte had about thirty slaves, brought by his sons, to farm the land he had been given. But most of them escaped; no more than three remained here. Many of the slaves from Samori's day escaped.

We haven't a single slave left now. They're all gone. The *tubab* came and said: 'Slavery's finished. *Allez!*' We're here on our own now, with our brothers, with our children and grandchildren. There are no slaves here. That was a long time ago.

Al-Haji Jaaje Jaaxo

Alimaami Samori, a ya ñi komon ragana; Alimaami Samori used to capture slaves. When he destroyed a town, he would take people as slaves, with their children. The Tanjigoora went with him, the Daraame went with him, the Jallo went with him; when he destroyed a town, they would take the women and children to sell. They bought them with cloth, with salt, and with grain for the troops' rations.

Then that was forbidden. The *tubab* gained power. They said: 'No one should buy or sell a person. Only savages do that.' The *moodini* who had bought slaves from Samori brought them back here. They set them to work, farming and chopping wood. They built large enclosures to keep them in. The slaves worked for them from early in the morning; they were alone, they had no one here. The masters rested. If a master was cruel, the slave would run away. Those who ran away never came back. Those who did not run away stayed here and had children and grandchildren.

Al-Haji Ma Jaaxon Ba

Laada toxo komon d'i kaman naxa. Between slaves and masters, only custom remains. There is no longer any slavery, any working for others.

Jaabe So

My great-grandfather Baraka wrote in his *taarixu* about people who were good to their slaves, starting with his own father, Bulel Asiya. Bulel Asiya was against slavery. He and his slaves shared the morning meal and the evening meal. When people said he was spoiling the slaves, he gave his slaves his cattle to herd, so that people would leave him in peace. Mammadu Denba was the same, and so was my father; in the end, the only slaves left were old women, who lived with us until they died.

In the days of Alimaami Samori, wherever Alimaami Samori went, Sanba Haawa would also go, with his horse. He bought slaves and brought them here. In the end, his slaves all escaped. He would leave them in the *jeeri* on their own; when he went there, he would find two or three of them gone. In the end he had only one slave left, named Kamele Njon, who was brought here young, grew up here, and stayed. He took care of Sanba's horse, fed him and watered him. Sanba gave him a house and found him a wife. Kamele Njon was the father of Siliman Sanqare's wife, Maama-lenme Jarra.

Sometimes fine horsemen were taken as slaves. Sanba Haawa bought one such slave, who said: 'Send us to Gece, to farm your land.' The slaves went to Gece, with grain and women slaves. When they arrived there, this slave gathered all the others together and said: 'When Sanba Haawa comes tomorrow, let's kill him; then we can all go free.'

A woman from Maama-lenme's family went to tell Sanba Haawa. When he arrived the next day, he said to the slave: 'If you like, we can fight, the two of us. But don't bring the others into it; I might lose my temper, and do them harm.'

The slave went as far as where the road from Bakel to the airfield is now; he sent another slave to tell Sanba Haawa he was waiting for him. Sanba went there, and killed them both.

Siliman Sanqare

Tanjigoorani do komon giri Mali; the Tanjigoora brought slaves with them from Mali. Some came here long before Samori's time; others came in Samori's day. Many of us know where their people came from. I know that my great-grandfather, my grandfather's father, was from Logo, beyond Xaayi. My father was from Wasulu. My great-grandmother, my grandmother's mother, was from Fuuta Toro, from—I've forgotten the name, the *jawando* town—from Senopalel.

Maama-lenme Jarra

N maama, n xooxo, n samba-laqe, my grandmother, my great-grandmother, my ancestor all lived here. It was Foode Mammadu Arjana who bought our ancestor, a long time ago. Her name was Sira Kante. She could not pound grain, nor fetch water from the river, nor cook. But a slave of Foode Mammadu Arjana said he wanted her. He bought her, and another slave to work for her. Her eldest daughter was called Manso Sira. Manso bore Boyi Manso, Boyi bore Laxane Boyi, Laxane bore Maama Xunba Laxane, Xunba bore Mayimuna Xunba, and Mayimuna Xunba bore me. Seven generations.

My father came from the East. The *cercle* was called Gudumi, and the town Ñeriko. In those days there was war. A man from here went out there; his name was Sanba Sire, they called him Sanba Haawa. His mother's name was Haawa Jinka. It was he who took my father and brought him here, with a rope round his neck. He made him his slave. My father had no one here born of the same father, nor of the same mother. But my mother was from here, one hundred per cent.

At the battle fought between here and Bakel, my grandfather's back was broken. A bullet struck him in the small of the back. They had been to destroy Senedebu, Bulebone, and other towns in Bundu. Sanba Haawa was a warrior. Muusa Kaba was a warrior; so were Biran Tuure and Baare Faatuma. Our grandfathers took up guns; they fought hard.

Siliman Sanqare

Komaaxu ñeme o faabani, slavery ended with our fathers. There was no quarrel, no fighting; we just parted company with the masters, that's all.

In the old days, they went to get slaves in the East and bring them back here. When they arrived here, the slave had no choice but to work for his master. His master would find him a woman to marry; the children belonged to the woman's master. The slave worked in his master's fields. If he went abroad and brought something back, he would give his master the keys to his trunk, saying: 'Here, take what you like.' The slave's food, even his clothing, came from his master.

Afterwards, many slaves left; some went back home, others went elsewhere, some established villages near Bakel. When those who stayed here began to have families, they started working part of the time for the master, part of the time for themselves. A slave would work mornings in his master's field, afternoons in his own. Monday and Friday were also his own. A slave's young sons would work mornings in their father's field, afternoons in their own small fields. That happened after *quatorze-dix-huit*. That was when slaves began to provide for their families themselves.

Now slavery is finished. *A toxon wa yi*, the name remains, but it's finished. No one farms for anyone else. No one does any work for anyone else. It's finished. You find a wife for yourself, you feed your household yourself; whatever you earn is your own. You may give your master a present, a robe or a pair of trousers; that's all.

What there still is between us, is that if there's a death in your former master's household, or a traveller returns, you'll help him to fetch firewood, and your wives will go there to pound grain and cook. After two or three days, they'll thank you and give you money. It's the same if there's a death in your own family, or a traveller come home: the women of your master's household will come to your house to do the cooking, and you'll give them money. That's how it is between my household and that of Mpali Hawa Tanjigoora. It's become the custom between us.

When slaves stopped working for their masters, some were old and didn't want to leave them. Moxobere, Galle Kanote's father, stayed with his master until he died. He kept on working in his master's field; he didn't have to, he just didn't want to change. There were others: like Bundukuru; like Sanba Jallo, who kept on working for his master until his master asked him not to; like Sanba Siise. Everyone else stopped.

When slavery came to an end, many slaves who hadn't the same name had become like sons of one father and one mother. When they were free, they joined together to form a household. Like Fode Jallo, Kula Jexite, Bubu Dabo, and Lasana Ndaw: they're not sons of the same father, but they united in a single household. Like Sanba Jallo, Laaji Sisoxo, and Baxayiro; like me and Laaji Tarawore. Or Galle Kanote and Haruna Kanote; they're not related, although they have the same name, but they formed a single household.

What we now call *komon golle*, 'slaves' work', is a neighbourhood society founded by slaves, those of Silimana and those of Tanjankunda. It was created a long time ago, for setting straw roofs on houses; you'd work the straw yourself, but when the time came, the society would come and help you set the roof in its place. Afterwards, when the river flooded and houses collapsed, people would help each other with making bricks and rebuilding their houses. The society was founded by slaves, but other people joined as well. Jaabe So belongs to it, because whenever the slaves wanted to do something, they would hold a meeting at his grandfather's. There are Tanjigoora in the society as well; there are *hooro* in the society, but it's the *komo* who give the orders. The society still functions: if there's a fire, they'll help rebuild the house. If someone who isn't in the society wants bricks, he pays for them, and the society keeps the money. If someone has a heavy job of work to do, he pays the society to come and help. On feast-days we buy cattle, and the meat is shared out among the members of the society.

Taalibo

Adrian Adams

The 'year before the call-up', Isa Budu was about twenty years old. He did not farm the family fields with his father and brothers; he was studying with Fode Mammadu, and worked with the other *taalibo* on Fode Mammadu's large field at Papata.

Isa Budu

Ken siine, that year there were about four hundred *taalibo*, a hundred to two hundred *xaralenmu* or younger pupils, and ten to fifteen *komo*, all farming for Xoje. The *xaralenmu*, the *komo* and the people of the household first worked for Fode Mammadu, then worked for each of his brothers in turn, starting with the eldest. The *taalibo* worked only for Fode Mammadu. They worked two days a week, Saturday and Sunday, from early in the morning to one o'clock. Monday was a day of rest. On Tuesdays, Wednesdays, Thursdays, and Fridays, the *taalibo* grew groundnuts on the sandy soil near Bema and Seenu. They sold the groundnuts in Bakel to buy clothing.

My elder brother and I harvested eight hundred *muudu*, maybe a thousand: about three tons. With the others, from time to time we would spend a day helping our *jaatigi*. My *jaatigi* was Al-Haji Abdullayi Tanbadu. I took all my meals at his house. We slept at Komonkani, at the furthest end of Tanjankunda; that's where our quarters were.

That year, Fode Mammadu's field was very, very large. The *jakka* alone was a hundred donkey-loads. He gave grain to the poor; he gave grain to his brothers if they needed it.

Al-Haaji Jaaje Jaaxo

Fode Mammadun kiye, in Fode Mammadu's day there were two hundred and eighty *taalibo*. They farmed for Fode Mammadu. If anyone was short of labour, you had only to ask, and he would help you out. The *xaralenmu* worked for the whole family, not just for Fode Mammadu. At Fanqanne, at Sanba-Salu, in the *jeeri*, the *xaralenmu* worked with Fode Mammadu's younger brother, Al-Haji Al-Xaali; they farmed a field for Fode Mammadu, a field for Al-Haji Al-Xaali, a field for Fode Alimaami, a field for Musa Jaaxo and so on; for all the brothers, ranked according to age.

I remember when we went to thresh Fode Mammadu's grain. Grain for the *jakka*, the tenth part of the harvest every Muslim must give as alms, was brought unthreshed into town. The *jakka* alone amounted to forty-five donkey-loads; so the harvest would have been four hundred and fifty donkey-loads.

Some of the *taalibo* were from here, or Golomi, or Yafera; many were Soninke from east of here, Jaafunu, Xanyaga, Soroma, Kingi; or from Gambia.

Sanba Jaaxo

Before the war, there were very few people who didn't farm in the *jeeri*. People farmed from Suxangide to Gaabu. I remember that in 1926 I farmed at Denba-Wuri, a hill near Gece, next to Fode Mammadu's field. It was an enormous field, like from here to Golomi. There were three hundred and fifty *taalibo* working there, not counting the *xaralenmu*. The *taalibo* also worked for themselves, in addition to the work they did for the *moodi*: they grew maize and groundnuts, which they sold in order to buy clothes and books.

Al-Haji Sanba Jimmera

Seexu Jomo renmen ya ni, Fode Mammadu was the son of Seexu Jomo. He studied with his father until God granted him wisdom. When his father was old, he took over his teaching duties.

Fode Mammadu had up to two hundred and twenty-five *taalibo* working for him. Many people started coming to consult him. He would send his visitors to stay with my father, Seyre Fenda. Every morning, the *taalibo* would bring *fonde* for the visitors' breakfast. At noon they would bring the visitors' midday meal to Seyre's house. In the evening they brought their evening meal. If there was a visitor of note at any household in the town, Fode Mammadu would present him with an ox or a sheep. Wherever the visitor came from, and however long he stayed, good food would be prepared for him, and the *taalibo* would deliver it.

When the *taalibo* harvested the household's grain, they threshed it and put it in the granaries; that's when the *jakka* was set aside. If there were a shortage of grain during the rainy season, Fode Mammadu would summon some *taalibo*, and tell them: 'Come back tonight.' In the middle of the night, he would fill sacks of grain and say: 'Give this to So-and-so.' The *taalibo* delivered the grain in secret. You could be in your house with nothing for the next day's meal, and in the morning find a sack of grain by your door.

Maama Kaba Tanjigoora

A yi debe su gotoono, Fode Mammadu persuaded the whole town to observe God's religion. When the month of fasting drew near, he would buy quantities of sugar and give it to the eldest of the women, to share out among everyone. People who never used to fast began fasting. Even little children fasted. Every year he would give sugar, saying: 'Let us follow God; we are not long for this world, but will go to judgement.' At Friday prayers, people from Jaagili would stand on the river bank to listen. In those days, there were many *taalibo* here.

Al-Haji Sanba Jimmera

A t'a ga na kara, Fode Mammadu said that when he died, his children would inherit only books; they would inherit neither money, nor cattle, nor donkeys, nor goats. Everything he received, he gave away. He told Seyre: 'If you hear that someone doesn't like you, be good to him. If you are good to him, perhaps he will change towards you; and even if he doesn't change, he won't be able to harm you.'

If there was a quarrel between townspeople, he would summon them secretly and talk to them, until they were reconciled. Anyone for whom he asked God's blessing would find that his affairs prospered. As soon as the rainy season ended, people would come to see him. Madmen would come: all the madmen of Gajaaga. All the madmen from Ñooro to Matam would come to Kuŋani. He would give a shirt to those who had no shirt, and trousers to those who had no trousers. The day he died, the madmen all gathered by his door, crying: '*O ma kara, o faaba kara!* Our mother has died, our father has died!' That is what Fode Mammadu was like.

When he died, his son Al-Haji Haamidu took over the teaching; but Al-Haji Al-Xaali took care of Friday prayers, and all of Xoje's affairs. Al-Haji Haamidu turned away from the things of this world. He gave away everything he was given; he never wore fine clothes. When he was not occupied with teaching the *taalibo*, he would go into his room and shut the door.

Women

Adrian Adams

Women stay at home, and they farm. The *jeeri* is their province; not the men's fields, where they rarely set foot, but the fields where they farm on their own. They do not grow the staple grain crops, nor build houses. But there would be no life in the houses without the water-jars they fill and the fires they keep alight. And the crops they grow, groundnuts, swamp rice, indigo, are necessary for life beyond survival: for enriching household meals, for porridge for the children, for dyeing fine hand-woven cloths. Men travel, but women stay behind, like the land.

Al-Haji Mpali Tanjigoora

O maani soxo; in the old days, our mothers farmed! They grew groundnuts and indigo and cotton; they did their own spinning. They also grew *fuñanŋe*. Our household began to grow *fuñanŋe* when they were living in Gambia.

Women worked hard in those days. Men didn't thresh the grain they harvested; every morning, the woman in charge of the day's cooking would thresh, winnow, and hull the grain herself. During the rainy season, she would then set it aside, and go off to her groundnut field. In the afternoon she would hurry back and pound the grain. When it was reduced to flour, she would prepare the *futo* for the evening meal. That wouldn't keep her from her groundnut and indigo fields the next day.

In those days, once the rainy season was over, a woman who was busy cooking would have her spinning basket at her side; she would stir the pot and spin. Every provident household head would grow his own cotton, to make thread for fishing-nets. He would say: 'Here is some cotton for fishing-nets. Share it out among yourselves.' The women would share it out; each one would spin her share, and give it to the *kagume*. He would give the thread to his sons, telling them to twist three strands together to make it strong, and work it into nets.

Our mothers worked very hard. All the clothing we wore was home-spun. There was a cloth called *kuubeli*; today's children don't know it. It was made of narrow strips of cloth; they spun and wove and dyed it themselves. You wouldn't have thought it was made of strips of cloth to look at it. We wore garments made of that cloth.

In those days, after the end of the rainy season, when there was less work to do, they would decide to have a 'girls' week'. That meant that for a week, seven days, the young girls would stay in one house, not going anywhere, not working for their mothers; just spinning. A hard-working person would twist the thread right after the week was over: some six cloths' worth, some seven.

The women grew their own indigo. They themselves went to burn wood for ash: the trees they used were *tunbe, waare, gese*. Women themselves would chop down the trees, split them, and burn them. Also, when men cleared land for farming in those days, they would cut down big trees; and during the hot dry season, they would burn the land they had cleared. Women would say: 'Hey! They're burning such-and-such a field today!' Late in the afternoon, at *laxaasara* or even *futuro*, they would go and collect the ash, to use in making indigo dye.

They used a *waate*, made by *tege* women: a clay pot with little holes. To process the ash, you put straw on the *waate*, with the ash on top. You added water until it was all soaked. You put something underneath, to catch the caustic substance it leached. Women used it for making soap, and for preparing indigo dye.

In those days, they harvested quantities of groundnuts. If you liked, you could take two or three donkey-loads to weigh and sell in Bakel. You could make soap too; in those days, there wasn't much *tubab* soap about.

Our mothers worked with all their strength. In those days, there were few households where the *kagume* supplied groundnuts. Since women grew groundnuts, they didn't look to the man for them; they themselves provided what they needed for cooking. They said that if you are with a man, each one should strive for the good of the household. If your husband works hard, you too, as a woman who farms, should show him that you had a good harvest of groundnuts. In the old days, women had that sort of pride.

When our mothers left Lonboli, they went to Sanba-Xonte, near the sandy soil of Baalu; when they left Sanba-Xonte, they went to Guñan. Then they left Guñan for Suxangide. That was when my mother stopped farming.

Haaja Sedinte Bacili

Ken bire, there was a time when it scarcely rained for seven years. The crops died before harvest time. We left our town, Segala, two days' journey upstream from Kuŋani, to go and farm in the *jeeri*, at Digi. We farmed there for seven years; we harvested enough to eat.

My elder brothers were both taken as soldiers for the Moroccan war. They both died. I was left the only child old enough to work. It was I who helped my father farm; I also worked with my mother.

I farmed with my father. I went with him to cut grass in the bush. When he made a fence, I held the wood upright while he made it fast. When the men went to cut strips of bark, it was I who carried the bark home. I worked with the men to prepare mud for building. In the fields, it was I who scared off the birds. I went to fetch hay for the horse.

Then my father died; we returned to the river, to Segala. I stayed with my grandmother. She was a woman who farmed for her family after her husband's death.

We farmed with our mothers. We grew groundnuts, we grew millet, we grew indigo, we grew cotton, we grew rice. It rained in those days. When we harvested our groundnuts, we would buy cattle in Bakel. My mother once sold enough groundnuts to buy a horse. We ate the millet. We used the indigo to dye cloths woven with the thread we spun from our own cotton.

People have always worked hard in Gajaaga. If you didn't farm, you wouldn't eat. If you didn't spin thread, you wouldn't have any clothes to wear. In those days, everyone had to work hard. Our skin never had time to cool.

In those days, male slaves worked for their masters; women slaves worked for their mistresses. I've never had slaves farm for me. I've always farmed myself. But pounding grain is something I've never been able to do. Mortar and pestle—no, that I can't do!

The Bacili are *tunkan-yugon renmu*, children of kings; but a Bacili who doesn't farm is living off the poor. I've always farmed. The poor have nothing; one can't just sit there. My grandfather was Tunka, but we will always farm. My parents filled their granaries from their own fields.

I first married a man from Gallaade. In Gallaade, the Bacili are *debegumu*. There are two Bacili households there; the rest are Gunjam and their *komo*. There's one Daraame household, who lead people in prayer.

I grew millet there. They have no groundnut soil; grain and indigo, that's what they grow. I grew a lot of grain. Women would sell their harvest to buy thread, then have it woven and sell the cloth. A woman who earned a great deal would buy gold for ear-rings.

After my first husband died, I married a *moodi*: Mammadu Tugaane, a younger brother of Sanba Koyita Tanjigoora. I went to Congo with him, and lived there for four years. He traded in cloth and taught the Koran. I didn't do anything; I just sat there, until my husband brought the food I was to cook. Out there, it's the men who go to market. I bore two sons there. Then we came back here.

It rained in those days. We farmed at Medina; we had good harvests. The first year, I had eight donkey-loads of groundnuts; the second year, twelve donkey-loads. I also grew *suna* millet, and harvested three donkey-loads. *Nke tanpi de!* I worked very hard.

In the morning, before setting out, we pounded grain. We swam across the *xoole*. We would put the small children in a big calabash, and talk to them while swimming, so they wouldn't be afraid. Women who didn't know how to swim would tie pieces of wood to their waists. At *sallifana*, in the early afternoon, the woman in charge of that evening's meal would gather a load of firewood and return home to cook, stopping at the *kolangal* to pick leaves for *dere*.

If the groundnuts were at a stage that required a lot of work, or the streams to be crossed on the way were running high, the other women would spend the night out in the fields. We built shelters ourselves, tying straw onto a framework of branches. We took *futo* and *ñecce*, and ate them with *dere* or with *taxaye*, a sauce thickened with powdered okra or baobab leaf. Jerebo, my co-wife, was good to me. She didn't spend the night out in the fields. She would say: 'You don't know how to pound grain properly. I'll do it for you.'

In those days, women slaves farmed for their mistresses in the morning; they farmed their own fields in the afternoon. They don't do that now, nor do the men.

Maama Kaba Tanjigoora

Soxoyen nt'a yi saasa wo! People don't farm nowadays. We used to work as hard as our strength could bear. When the rainy season came, each woman would pound her own *futo* and *ñecce*. In households where there were three women, or four, or five, they would take it in turns; one would do all the cooking for a week, while the others would stay out in the fields, at Jabalu, Bema, Medina, Sanba-Guro. We would stay out there until we'd finished sowing our groundnut fields. We would go back out into the fields when it was time for weeding. The fields were weeded twice; then we would take it in turns to guard them until harvest time. When the time came, the women of the house would go to help the men harvest their sorghum. Then the men would help bring in the groundnuts.

Some women would have ten donkey-loads, or twelve, or thirteen. In those days, men weren't expected to provide groundnuts for cooking. Each woman would provide her own groundnuts when it was her turn to cook, from the start of the year to its end. You would also sell some of your crop in order to buy cloth, or black and white thread to weave and dye with indigo for your daughter's bridal *kande*. Now they've given all that up. Women won't spend the night out in the bush; they sow little fields at Suxangide, where monkeys from the town can go and spoil them. Call that farming?

We used to farm from here to Hamdallayi, from here to beyond Gaabu, to Sanba-Xonte. We would forget our homes, and live under shelters; only when the groundnuts had been gathered would we all meet again in town.

We also grew indigo; some women would end up with twenty or thirty loaves of dried pounded indigo. We grew cotton and okra. In Jaagili, across the river, they were poor in those days. Their land wasn't as good as ours. They grew *mulunqu* in their *falo*; they would bring baskets full of *mulunqu*, to exchange for sorghum and groundnuts. They would go far out into the bush in search of *keye*. They were poor. But Kuŋani lacked nothing.

Everyone farmed, women and men. My father Muusa farmed with his *kome*, Sanba Tene's father, who had a large household; he always had a huge stack of sorghum at harvest time. He grew a large red variety of cowpea; all the women of the town would help him gather it.

In those days every woman laid by a supply of *futo mulle*. When there was good milk to be had, you would call your husband into your rooms at mid-morning and offer him *futo* and milk. You would cook *sonbi* in the afternoon. A woman would buy fish from the river, to add to her *dere*; or she would cook *futo fillante*, and say to her husband: 'Come and taste this small thing I've prepared for you.' A woman would advise her husband in private, and turn him away from bad decisions and unkind thoughts. Now it's all dried-up rice, and groundnuts bought with money from France. I know one thing, though it may find me in my grave: there'll be an end to the money from France.

3

Abroad and Home
(1937–1968)

Senegal

Jaabe So
Ken waxati, during the colonial period, the only forced labour we did here on the River was in our own interest: working on the roads so we'd be able to travel between towns. What else was there? Just paying tax. And things were judged fairly in those days.

One day we went to do forced labour on the road near Bakel, where the creek-bed runs through the Bakel *kolangal*. That was in 1935. A *gendarme* was there with us, a Fula. When we finished filling in the creek-bed, he tied up the bundle of wood he'd been collecting. People started leaving the *kolangal* to return to Kuŋani; the people from Bakel, Bema, and Guñan also started to leave.

The *gendarme* said to me: 'You boy, pick up that bundle of wood and take it to Bakel.' I said: 'No, I'm not going to Bakel; I'm going to Kuŋani.' He said: 'Take that wood to Bakel.' I said: 'No.' 'Take that wood to Bakel.' 'No.' He came and took hold of my arm; I wrenched free, and he slapped me. I hit him on the head with my short-handled hoe. He beat me and put me in chains, and made me walk in front of him all the way to Bakel. The bundle of wood he'd wanted me to take got left behind after all.

We went to the *préfecture*. He put me in a room, and locked me in; then he went away. At three-thirty the *préfet*, who was a *tubab*, looked in and saw me there. We couldn't understand each other; he sent for an interpreter, and told him to find out why a boy like me was locked up in there. I told him the whole story.

Now I come to why I say colonialism's day and our present day are not the same. The *préfet* had the *gendarme* summoned, and said he had no right to make me carry wood for him, let alone beat me. He rang up the *commandant* in Matam, and asked him to give orders for the *gendarme* to be called to Matam and dismissed from service.

That would never happen nowadays; they would gang together. There's no justice. If a civilian hit a *gendarme* nowadays, they'd beat the life out of you. There'd be no chance to tell the truth. The reason some of us were in favour of Mamadou Dia, was that we thought if he were in power, justice would prevail.

One day, it must have been in 1933, we had been doing forced labour on the road near the Bakel *kolangal*, and decided to return to Kuŋani by a short-cut through the bush. We were with a *kome* who knew many things. He called me aside and said: 'Do you see that rock? That is where Seega is buried.'

In 1936, at the time of year when the river floods, I left here with Bakari Kamara. We were the same age. When I left, I stole an indigo-dyed cloth from my mother; and I had a gold ear-ring in my left ear, fifteen grams of gold.

We decided to go to Dakar. There was the train in those days, and it didn't even cost much; but we didn't know. Instead of going to Kidira, we set out through Bundu: Bala, Kocari, Tambacounda. We sold the indigo-dyed cloth along the way; we had money, a calabash, a small packet of *futo*, and a loaf of *beke* sugar.

When we arrived in Tambacounda, we stopped there for a few days. I took a job chopping wood and washing dishes for a Fula woman named Ayisatu Ba. One day she called me to come and wash her back; I said: 'No, I don't want to.' She was angry and said I would have to leave. That evening her husband told me: 'My wife says you're too stubborn.' He gave me my four days' pay.

We kept going. We walked and walked; sometimes night would fall before we reached a village. If there was a moon, we'd keep going. If there was no moon, we'd stop, and sleep under a big tree. Bakari Kamara would sleep next to the tree, with me on the outside.

I sold my gold ear-ring in Guinginéo. We still had money left from selling the cloth, but I was afraid it would soon run out; we didn't know how far we had left to go. I sold it for only fifty francs, to a man at a stall selling candles, bread, and sugar. When I told my brother that, in Dakar . . .

It took us about thirty days to walk from Bakel to Dakar. When we arrived in Dakar, we found our way to the *konpe xoore*, the house where our townspeople lodged. We arrived there on 30 June 1936, at five o'clock in the afternoon.

My brother was working for a Lebanese shopkeeper. Bakari Kamara's brother was working for the Shell Company. In those days, there wasn't much money about. We spent about three months in Dakar. After two months, my brother wanted me to go home; but I refused. I wanted to spend the rainy season working in the groundnut-growing area.

I set out on the road from Dakar to Kaolack. Badara Geyi was working in Kaolack as a policeman; but I didn't want to go to his house, because he might decide to send me home. We spent the night in the market. In the morning, we made our way to the farming country nearby.

We went to Mpanesadere, near Kër Mejebel. When we arrived, they were hiring *surga*, seasonal workers. All the others found someone to work for, but I was too young, and no one would take me on. I bought a sack of groundnuts for seed; but I didn't have a *jaatigi* to stay with. The way things worked there, was that your *jaatigi* gave you land to farm; you worked four days a week for him, and three days for yourself. But I couldn't find anyone to hire me.

There was a man there named Birom Tik, who had many *surga* and many wives. He was a Wolof, a *lawbe*; he had twelve wives. He was like a king; twelve wives and many children. He said: 'Why hasn't So found anyone to stay with?' My companions said: 'It's because he's too young.' He said: 'Is that so? Well, I'll take him on.'

He hired me, along with other workers, and gave me a plot of land. I spent the rainy season in Mpanesadere. There were seventeen *surga* in his household; we farmed groundnuts throughout the rainy season. It was hard work, and I was sorry I'd come. The food was not good: their *futo* was coarse, a bit sour, and they ate it with a sauce made of the leaves of trees. But since I'd sown my groundnuts . . .

At the end of the rainy season, we harvested our groundnuts. I wasn't strong enough to harvest all of mine, but I kept on working. One day, Birom Tik asked if everyone had finished. They told him: 'So hasn't finished.' He told all seventeen *surga*: 'Go and harvest So's groundnuts.'

Once the groundnuts were harvested, they had to be threshed. The people who'd gone to work there formed groups, each *xabiila* together: Tukulor together, Soninke together, people from the East together. In those days, there were many Malians, not many Senegalese. There were eight of us there from Kuŋani, Sanba Ñama and the others. They wouldn't have me in their work group; they said I was too young. There was a group of Malians staying at Birom Tik's house; they said: 'So, you come with us.' We threshed all the groundnuts. When my groundnuts were threshed, I had six hundred-kilo sacks; more than Sanba Ñama. Only two of the people from Kuŋani harvested more. Birom loaded my groundnuts onto donkeys, and we went to Kër Mejebel to weigh them. I took my money. In those days, a hundred-kilo sack of groundnuts was worth only 400 francs; so I had earned 2,400 francs. I put 200 francs in my pocket, and tied the rest around my waist.

I went back to Kaolack, and went to see Badara Geyi. He told me off for going to farm groundnuts without telling him. I showed him my money; he said that he would have some clothes made for me, then send me home. I said: 'No, I'll just buy one robe, then I'll go to Dakar to see my brother Siixu. There's good cloth for sale there; I'll buy some, then go home.'

At the time when I arrived in Dakar, my brother wanted to go to France. There were seamen's *fascicules* for sale for 300 francs. My brother talked to a man, but in the end he said it cost too much. The man left. He was a Soninke; I followed him. I said: 'Can't you lower the price?' He said no; the price was 300 francs. I paid him; I put the *fascicule* in my pocket, and came back to where we were staying. We had our midday meal. I didn't know Dakar; I went out for a bit of a walk. Siixu had his eye on me: 'Don't go too far! Don't go too far!' That evening I spoke to Sanba Jaaxo, and told him: 'I bought the *fascicule* from the man who was here this morning.' He said: 'What! Do you mean you paid for it?' I said yes. I gave the *fascicule* to my brother. Then I had some clothes made, and went home. That was in 1937.

At the end of the 1937 farming season, I returned to Dakar, and worked as a labourer for the Shell Company, filling petrol drums. I lived at 94, rue Raffenel; that was the Kuŋani *konpe xoore*, a house rented by our townspeople. There were many of us living there; sometimes as many as thirty, all working in Dakar. Travellers arriving from Congo would stay there; travellers arriving from France would stay there. Everyone was welcome. There was a fund for expenses; everyone paid a contribution at the end of each month.

Adrian Adams

Twenty-five or so other Kuŋani men were also working in Dakar as labourers. Fewer than a quarter of them were *moodini*, and about three-quarters were from *komo* households, with a few *tunka-lenmu*. That same year, ten Kuŋani men were working as seasonal labourers in the groundnut basin of western Senegal, as Jaabe So had done two years before. Twenty-odd men from Kuŋani were working on ships of the French merchant marine: including Sanba Haawa's son Npamara Sanba Tanjigoora, MaJaaxo Tanjigoora from Fode Sire's household, Kuliko Gunjam, Siliman Jabira, Denba Xulle Jabira, and Jaabe So's brother Siixu. Denba Xulle was using Bakari So's papers.

War

Jaabe So

Ken siine, in 1939 I went to the Dakar naval base for a medical examination. They said I was too young. They said: '*Petit*, go away, come back next year.' So I left. But I was determined to join the Navy; I managed to have some papers faked, to show that I was twenty years old. A month later I went back there with my counterfeit papers, and was accepted.

We used to say Gajaaga was a pleasant country, because nothing was lacking here: there was grain, and meat, and fish in plenty in the river. People were people then: we cared about each other, we helped each other.

There were horsemen and wars in Gajaaga; then the wars came to an end. The French had an army, and we joined that instead. All fine men here were warriors; so we volunteered for the army.

We didn't join for the money. In those days they really took advantage of us, a *soldat première classe* was paid twenty-five francs a month. In the Navy, on board the submarine supply vessel *Jules Verne*, on board the *Liberté*, the *Montcalm*, the *Georges Elek*, or the *Jeanne d'Arc*, seamen were paid twenty-five francs a month, a hundred francs if you were a corporal. People weren't looking for money. They were looking for dignity as men; they wanted to do as fine men had done in the past, to see battle and win distinction. It was hard sometimes, but men should endure hardship.

That was why we left our country to join the military. The horsemen of former times earned their name in battle. Now the Bacili don't fight any more, and battles have vanished from our land; we went looking for battles. I saw many men die at the landing at Bizerte, who were paid twenty-five francs a month. They weren't looking for money.

Abroad and Home (1937–1968)

I did all my training at the Dakar naval base. Then I was taken on board the *Jules Verne*, a submarine supply vessel. The war was still on.

My brother Siixu was in the merchant marine; he left France in early 1940, and came to Dakar, then to Kuŋani. When his train stopped in Tambacounda, Muusa Gayi Tanbadu was standing on the platform; they greeted each other.

At that time the French were against De Gaulle. Muusa Gayi's elder brother, who worked in Dakar, had left a De Gaulle leaflet with him. When it was found and he was questioned, he said he had it from Siixu So. They sent a message to Bakel. The *commandant* there was a bad man.

I myself was in Kuŋani then, on leave before our ship departed; Siixu and I slept in the same room, but he rose very early to go and see Al-Haji Buna Tanjigoora in Golomi. I set out for Golomi after breakfast, to see my friend Haruna Tanjigoora. As I was entering Golomi, the *commandant*'s car drove past me. They went to the *moodi*'s house to take Siixu into custody. The *moodi* told the *commandant*: 'That car won't reach Kuŋani.' The car broke down, and they entered Kuŋani on foot. They put a guard at the door of our house, and started searching our belongings. When I arrived, the *gendarme* told me no one was allowed in. I was in my sailor's uniform. I said: 'I'm going in; those things you're searching are mine as well.'

The mean *commandant* was standing in the courtyard. He said: 'Don't come in. If you do, I'll order them to shoot.' I said: 'Go ahead. I'm coming in; this is my house, and those are my things. Those are my uniforms in that suitcase. Siixu's not in the Navy.' The whole town came in and sat down in the courtyard. The *commandant* left, taking Siixu with him.

I went to Bakel after them. Our mother gave me ten thousand francs to give Siixu. He said he had enough money; let her just pray for him. I went to Dakar straight away; I went to see Lamine Guèye, who was a lawyer, and told him they were going to kill my brother. Lamine said to come back on Monday. On Monday, he said: 'The day after tomorrow, they're sending your brother to Bakel. They've released him.' I owed him three thousand francs, that was what you paid a lawyer; I told him I hadn't any money, but I would pay him as soon as I could. When Siixu arrived in Bakel, they drafted him into the Army.

As soon as I'd managed to save up a bit of money, I went back to see Lamine Guèye. I'll always remember that day. He said: '*Ah! C'est le petit So?* What was this morning's parade all about? I said: 'Some admiral came off a ship to visit the naval barracks.' When I gave him the money, he held it in his hand and smiled. He said: 'How much are you paid?' I said twenty-five francs a month. He said: 'How did you earn this money?' I said: 'People pay me to wash their clothes.' He didn't say anything for a while. Then he said: 'I want you to take this money back.' I refused. When he insisted, I said: 'That money is *haram* to me.' I put it on a table and left.

My mother's two sons, Siixu and I, were both at war; she was beside herself with worry. At that time the *tubab* were going about taking people's cattle and sheep, saying it was for the war. One day the *commandant* came to Kuŋani. My mother had a fine fat ram; the *commandant* told a *gendarme* to take it. My mother said to the *gendarme*: 'Do you mean to say you're taking that sheep?' He was a Fula. He said: '*Ko geeri*; it's to feed the soldiers.' He had the ram by the horn; my mother took hold of its leg, and said: 'Wait a minute; this sheep belongs to me.' People came to see what was going on, and she told them to call the *debegume*. When he came, she said to the *commandant*: 'My two sons are at war, I don't know if they're alive or dead; now you say you're going to take my sheep! This sheep is not going anywhere.' The *debegume* told the *commandant*: 'Even if you kill that woman, you won't take that sheep away.'

The *commandant* left. As he was leaving, my mother said to him: 'You're a *tubab*. The Germans have attacked France, and you've taken my two sons away to die. Now you're going about taking people's livestock. Have you no fear of God? Why aren't you at war yourself? You ought to go; it's your country.'

The *Jules Verne* supplied food and water to French and English submarines. We were always at sea, sometimes for as long as two months at a stretch. All the submarines in the Atlantic would come and get food and drink in mid-ocean.

The war continued, on that same *Jules Verne*. We were told to go to Algiers. We left Dakar; between Dakar and Port-Etienne we were attacked by a German submarine. It fired a torpedo, but missed. The captain of the *Jules Verne* went into the harbour at Port-Etienne. He telegraphed Dakar; some small boats came up from Dakar with nets to close off the harbour, so that the German submarine could not destroy the *Jules Verne*.

All the ships were in harbour; because of the nets, the submarine could not get in. It was a German submarine, supplied from Spain. For three months, we couldn't go on to Algiers. That submarine sank many ships between Dakar and Casablanca. Finally, a little *chercheur-torpilleur* came from Dakar to search for the submarine, and sank it. Once the submarine had been sunk, the net was removed and we continued on our way to Algiers.

We arrived in Casablanca and docked there. I heard that my brother Siixu was there as well; their convoy had come from Dakar to prepare for the landing at Bizerte. That evening, I asked for leave to go and see him. When I got back, the *Jules Verne* had already left for Algiers. I went to the naval base to say that I'd been left behind. A submarine, the *Surcouf*, was due to go to Algiers the next day; I went with them, to join the *Jules Verne*. We spent almost two weeks in Algiers, preparing for the landing at Bizerte.

I heard that the bad *commandant* who used to be in Bakel was in prison at Maison-Blanche. One Saturday, I went there with a group of sailors. I was with a young man named Aka, from Côte d'Ivoire. When we came to where the guards were, we saw a *tubab* in white shorts, behind a wire fence that was electrified at night. An Algerian told us: 'That's the *commandant* from Bakel.' As soon as we drew near, I recognized him.

He came over and spoke to us. He asked Aka where he was from. Then he asked me. I said: 'The Bakel area.' He asked: 'Bakel itself?' I said: 'No, from the *jeeri*.' He began: 'Guña, Bema . . .' He named all the *jeeri* villages. I said: 'You know them all.' He said: 'Yes, I am *commandant* So-and-so, I was there for a time.' I said: 'Don't you recognize me?' He said no. I said: 'I recognize you. Do you remember the day you went to arrest Siixu So in Golomi?' He said yes. I said: 'Do you remember the sailor who came after him?' He said yes. I said: 'Afterwards, when Siixu came back, you drafted him.' He just looked at me without speaking, then looked down.

'You made the two of us go to war. I volunteered; but Siixu was drafted. Then you went after our mother as well. Didn't you have words with a woman in Kuŋani? I'm from Kuŋani.' He said: 'Ah, yes. *Elle était tellement méchante* . . . I said: '*C'est vous qui êtes méchant.* You're paying now for all the evil you did then.' He walked away. We saw him go and sit on a bench. Afterwards, I caught a glimpse of him again, in prison in Algiers. I've forgotten his name. I asked about him afterwards in Toulon, but I never heard anything more.

They put all the warships on alert: the *Jules Verne*, the *Montcalm*, the *Jeanne d'Arc*, the *Emile Bertin*, and all the American ships that were there. Apparently the Americans had landed at Enfidaville, and it hadn't worked out; the French were to conduct the landing, with American landing-craft and French cruisers.

They asked for people to put down their names as volunteers for the landing. I put down my name; the purser said no, but I insisted. So I ended up as a *fusilier marin*, on board the *Emile Bertin*.

I had a friend, Sili Jeremuna Sumaare; he was from Hayire, downstream from Gajaaga. Wherever I went, he would follow. When I went on board the *Emile Bertin*, he also volunteered to leave the *Jules Verne* and join the marines.

One evening we were together on board the *Emile Bertin*, in Algiers. Sili's family's praise-singer, a merchant seaman named Denba Ñelle, heard that Sili was there. Everyone knew Denba Ñelle. His ship, the *Ville de Massanga*, had nearly been hit by a torpedo, and the men were afraid to go down into the hold. The *commandant* told Denba to sing to the men to give them courage as they loaded the furnace with coal. Once the furnace was lit, they would take it in turns to stoke the fire, two by two. From then on, from the start of a voyage to its end, Denba's only work was to stand at the mouth of the furnace and sing.

He came into the harbour and stood next to the ship. Then he called out Sili's name: 'Sili Jeremuna!' Sili said: 'Denba! Come on board!' Denba said: 'That's not what I've come for.'

Denba began to sing. He sang of things that happened in our country in the old days: of Sili's grandfather Mammadu Denba Jeremuna, a fine horseman; of Mammadu Denba's grandsons, two handsome boys who died in the Dardanelles. He sang and called out Sili's name. Sili was there on board the *Emile Bertin*, weeping. He wept so much . . .

Everyone on board, the captain, the lieutenant, the seamen, came to listen. The captain asked: 'What's going on?' We said: 'It's his *griot*.' In those days, a sailor had nothing. Sili asked the captain to let Denba come on board. He came aboard, and spent the evening with us. We left the next day.

We landed at Enfidaville. It was very hard, with many dead; but we managed to land all the same. There were over two thousand soldiers in the first landing; the Germans killed them all. In the end, they put the marines and soldiers together. The next day, the *Emile Bertin* went to fire on Bizerte. The Germans began to retreat further into the countryside. Patrols went out every day, soldiers and marines separately. The patrols began to go beyond the town, out into the bush, to see if there were any Germans left there.

One day, we went out on patrol with Lieutenant Roger: Sako Usuman and Sili Jeremuna from Hayire; Njayi Xaramoxo from near Kaolack; So Idrissa and Umar Ja from Fuuta. Out in the bush, we went into a deep ditch. There was a German with a machine-gun behind a big tree. He fired on us, and So Idrissa fell. I saw the German behind his tree; I wanted to run towards him with my automatic rifle. Sako Usuman took hold of my feet, and dragged me back into the ditch. He said: 'Do you want to get killed? Stay here.'

We stayed there and waited for the German to come out. He was keeping an eye on us. Sako Usuman crawled a long way round, until he came up behind the German. Sako didn't have an automatic rifle, he just had a bayonet on the barrel of his rifle. When the German sensed there was someone behind him, he started to turn around; Sako thrust the bayonet into his back. He cried out: 'Aaaa . . .' Sako said: 'If he's the only one out here, he's dead.' We all came out of the ditch.

That was when we saw that So Idrissa was dead. Bullets had gone through his chest. The lieutenant said we should leave him in the ditch, with a bit of earth over him. I said to Sako: 'What? Leave So Idrissa in the ditch? We've got to take him back to camp.' At that time, there were dead bodies all over the bush; but I kept insisting: 'We must bury him.' Finally, the lieutenant agreed, because I said: 'If he stays here, we'll all stay here.' We took him back to camp and buried him.

After the Enfidaville landing, we had German prisoners. Every morning I went to fetch the bread, with a driver and a prisoner. One day I saw the German pick up a crust of bread that had fallen to the ground, and eat it. I gave him some bread. Afterwards, the *second-maître*, whose name was Bébert, told the lieutenant: 'So gives the German prisoners food.' So they put me in prison too.

The admiral said to me: 'Didn't you see all the people the Germans killed; all your brothers they killed? You've no right to give Germans food.' I said: 'The Germans killed my brothers; my brothers also killed many Germans. I can't bear to see people go hungry. You can kill a man; but if he's hungry, you must give him food.' I was sent back to prison; a bit later, they let me go back to work.

After the Enfidaville landing, after Bizerte, after Toulon, we went to Cherbourg. At Cherbourg I caught cold, and fell ill. I was sent to Toulon in an aeroplane; I was kept in the Sainte-Anne hospital until the end of the war.

My brother Siixu had been studying while he was in France. Everyone was amazed at how much he'd learnt. He began working in a hospital. In 1943 he passed an examination in Casablanca, along with three Bambara and two Fula. The Bambara all had their *certificat d'études*. Siixu passed the examination and became *infirmier d'Etat*. The Bambara were not pleased at that; and one of them was a caster of spells. Siixu was on night duty when the spell was cast upon him; he suddenly saw something, and cried out. From then on, his mind was deranged.

No one knew what to do. There was a colonel named Delacoste, who took care of Siixu and had him put in hospital; when they decided he should be treated here, Delacoste brought Siixu to Dakar by aeroplane. At that time, the Toulon *débarquement* was being planned, but had not yet taken place. I wrote to the colonel asking for Siixu to be removed from hospital and brought here. The colonel and a lieutenant brought him to the *préfecture*; then they and the *préfet* brought Siixu to my mother in Kuŋani. The colonel sent fifteen thousand francs to help treat him.

Siixu remained here in Kuŋani until he died. He was very quiet; he knew people, but preferred not to talk to them, or go anywhere. My mother grieved so much that it hardened her heart.

After the war, we came to Dakar. I asked for twenty days' leave, and cam home with some other men. All returned soldiers and sailors were summoned to the Armistice Day celebrations in Bakel. The *chefs de canton* were there: Lieutenant Saada Sire Timmera from Yafera, and Ibrahima Jaaman Bacili from Tiyaabu. There was also a *tubab*, the *commandant* of the Bakel fort.

I got there late, and Saada started telling me off. I told him: 'Look, I've seen more than your parade. I've not come back from there to be told off by a *chef de canton*.' He asked where I was from. I told him: 'Kuŋani.' 'What's your father's name?' I said: 'Amma So.'

Bacili said: 'Saada, this man's not yours; he's mine. He belongs to Tiyaabu.' He took me to his house to eat. He said: 'I'm going to ask the Navy to demobilize you immediately. You'll stay here and help me. You can marry any woman you choose; and I'll buy you any horse you like. Your family have always been organizers for the Bacili; I can trust you.' I said: 'Wait until I talk to my mother.'

I went home and spoke to Juumu, my father's first wife. Juumu said: 'You must refuse. Everyone is against Mallalu Jaaman; if you work with him, everyone will be against you too.' So, even though Tiyaabu is like

> In 1944 this man of about fifty, a schoolteacher and *ancien combattant* of 1914–18, agreed to assist his uncle, Tonka Konko Golla. Of French culture, a member of the SFIO, he broke off all connection with the members of the Bathily clan, whom he considered to be backward; opposition to his authority became general, among the Bathily, the marabouts, and even among newly demobbed *tirailleurs*. Despairing, he shot himself on 26 June 1947, at Tuabou. 'He was not even acknowledged as being ahead of his time; the population showed only surprise.'
>
> (J. Robin, 'Autour d'un suicide sénégalais', Conferencia International dos Africanistas Ocidentais, *Bissau, 1947*)

our right hand, I didn't go to Tiyaabu. I was demobilized in Dakar, and joined the merchant marine.

In 1946 I married Maamu Sakiliba, the daughter of Waali Silaamaxa Sisoxo. She was born in Kulikoro; her father was a mechanic on the railway. I'd first met her in Bakel at the end of the war, then saw her again in Dakar where she was staying with her sister. We were without children for a long time. When our son Mammadu was born in 1958, my brother Siixu was in hospital in Dakar; I went to see him, and took Mammadu with me. Siixu said: 'Jaabe, so we too have a child now?' He held the child on his knee, and wept. Then he said: 'You can take him away now.' Soon afterwards he went back to Kuŋani, and died there.

On board ship, I used to read the book where my great-grandfather Baraka had written his *taarixu*. Siixu Jagana, a man from Wawunde, would read it to me. Later on, Salun Geyi asked me to lend it to him. I was boarding ship then; I left the book with Salun, so that my brother Siixu could read it to him. Then Salun died. I asked people to search his belongings, but the book was nowhere to be found. When Juumu heard what had happened, she was very angry. She said: 'Your father had two sons, Siixu and you; but he meant that book for you. He said so before he died.'

I myself grieved at having lost the book. But if it had remained with me, it could have made trouble.

Merchant Marine

Jaabe So

Ken siine, in 1946 I joined the merchant marine. I started work with the Compagnie Worms, on the Madagascar line. The first ship I worked on was the *Ville de Massanga*, a *cargo mixte* taking wine to Madagascar. Two years later, I signed on with Messageries Maritimes. We went to New Caledonia, Tahiti, Japan, sometimes Russia, and South America; my knowledge grew.

In those days, there was a lot of racism. One day in Tahiti we went to a bar, and they said black men weren't allowed in. I went in. There was a *tubab* there, a Frenchman; he said '*Mon petit*, no blacks in here.' I said: 'I never heard of black men not being allowed in bars in Tahiti; isn't it a black man's country?' He said: '*Ne fais pas d'histoires*; get out.' He came over to take hold of me, to throw me out.

I said: 'Don't do that. Say what you like, but keep your hands off me. It's your bar, you don't want to serve me, *d'accord*; but I won't let you take hold of me.' Another *tubab* who was in the bar told the owner to leave me alone. 'Those are Senegalese,' he said. '*Ce sont des Sénégalais*. I know them.' So I left quietly. We went back on board and told people there what had happened.

In New Caledonia, Maxan Sire had a lover in one of the bars; a fine girl, a beautiful girl. The bartender, a *tubab*, hired three black men to follow Maxan down to the shore and do for him. Maxan spent the evening in the bar; when he left, the girl went out with him to say goodnight. When they'd parted, he began to walk down to the harbour. Three youths were lying in wait for him; they stepped into his path, holding knives. They attacked him, but their knives couldn't touch him.

Maxan had a pistol; he shot one of them, who fell. He knocked down the others; then he took out his pistol and made one of them remove the other's belt and tie his arms with it. When he'd finished, Maxan himself took that man's belt out and tied his arms with it. Then he said: 'Let's go down to the harbour.'

The man he shot died, and the police arrested him. They said the ship should not be allowed to leave. But the captain came to an agreement with the local people. Messageries Maritimes said the ship should be allowed to leave, because Maxan had acted in self-defence; the men had been hired by the *tubab* to attack him. The company's people in New Caledonia would see to it that justice was done.

Maxan Sire went back to marry the girl and brought her to Marseilles. But she caught tuberculosis and died.

There wasn't a woman like her in all Marseilles. When she went out, people would just stand and look at her. God made their women unlike any other women. Anyone who has seen them knows that.

I spent five years with Messageries Maritimes. Afterwards, I was tired, and I wanted to see more of my family; I signed on with Freyssinet and Fabre. They did the coast of Africa: Dakar, Conakry, Abidjan, Lomé, Cotonou, Takoradi, Lagos, Pointe-Noire. From time to time a cruise, to Portugal or the islands.

I liked working on ships. In those days a cook was paid 60,000 francs a month; that was good money. But all that time, I wasn't just after money; I was after knowledge as well. I was looking for knowledge of the world. In every port where there was farming country nearby, I'd take a taxi after work and go and look at the fields. What I really liked was the maize-fields in South America. The first country we called at, there was a great bay: Rio. The first few times, I didn't even think of going ashore; but we often called there. Once when we had engine trouble, we spent over a month there. I met a Brazilian who spoke a little French, like us. We became friends; whenever we arrived, he would know it was my ship, and come on board with his big hat.

I used to ask him questions. When I saw their beef, I said: 'Are those your cattle? So big that you can't hold two legs at the same time?' He said yes. I said: 'I'd like to see them alive.' He said: 'There's a lot of cattle in the countryside here.' We took a taxi and went out there. I saw big oxen, twice as big as ours. A bit further along we saw them ploughing maize-fields. That was where I first saw oxen ploughing. One time there was no maize, it had all been harvested; but the stalks were left, and they were ploughing them under.

Every time we stopped in Rio, I went out to the countryside. That really pleased me. I thought to myself that in later life, if I lived a long time, I would like to take up farming again. That was when I first had the idea of coming back here when I retired, instead of staying in Dakar.

The Seamen's Union decided to recruit more members; they recruited me. I went all over, to Bordeaux, Le Havre, and Dunkirk as well as Marseilles. In those days the shipping companies were robbing us, they paid only three hours' overtime for jobs done mainly by black men.

One day the union printed papers saying no ship should leave harbour. They gave me those papers and I put them up everywhere, on all the ships and other places where people would see them. We went on strike. Five days later, overtime went up to four hours, five hours, six hours, seven hours: the same as for white people. Before, black men weren't entitled to a pension; afterwards, they had the same rights as white men.

All the time I was working on shipboard, I worked with Corsicans; we got to know each other. In Marseilles, I played poker in their bars. Mémé Guérini used to play in the Bar Cloche. One evening when I won a lot of money, there were some Corsicans from Toulon, who didn't know me; they said in their language: 'Let's not let that Black get away with all our money.' The woman who owned the bar was afraid; she told me to wait, and she'd take me home in her car. I said no. I put the money in my pocket, and made sure my pistol was in my belt; then I put on my jacket and left. I walked all the way from the Bar Cloche to the hotel I owned then, 44 rue Nationale; I stood still for a bit at the Grand Place, but I didn't see anyone. I heard afterwards that Mémé Guérini had told the others: 'Leave that man alone; otherwise there'll be something in the papers tomorrow that will shame us all.'

We knew each other. One day they rang up Mémé Guérini from London, to say that a man who'd been expelled from France, Roger André, would be arriving at the Marseilles airport. The Corsicans rang me up to ask me to reserve rooms for them. Somehow the police found out. At that time there was a Commissaire de Police in Marseilles named Albertini; he knew me well. They rang to tell me not to receive those people in my hotel. I told them that I'd already agreed; I couldn't go back on my word now. They weren't pleased; they said; 'Whatever happens to you, you'll have brought it on yourself. That man is *interdit de séjour*. When he arrives, we'll be there.'

He was met on arrival by six black cars; they brought him to the hotel. I gave them Room 6. He was with his wife, a short woman. Shortly before ten, *Sécurité* police cars drew up in front of the hotel. They asked me for the room number, and I gave it to them. They went and knocked on the door. Roger André said: 'I'll open the door. You know I could make it tough for you. But I'll open the door for Jaabe So's sake.'

He opened the door. The police went in; they said: 'Hands up', and he raised his hands. They said 'Hands up' to his wife, and she raised her hands. The wife had a belt round her waist; they unfastened it. They unfastened his own belt as well. Both of them had an Adams pistol. Roger André called my name. He said: 'They're not tough enough for me. But I'm surrendering so you won't have any problems.'

My hotel was *signalé*, but I had more friends than the police lawyer, so it wasn't too serious.

I made friends in France. Madame Claire was an Italian woman; she ran a *blanchisserie* rue des Dominicaines, where numbers 10–13 are next to a church, so it's like a little square. When I was new in Marseilles, I passed by there every day, and she would greet me. One day she called me into her shop, and asked if I had been in Marseilles long. I said no. She gave me a shirt, and told me to bring her my clothes to wash. She had no children, and I became like a son to her; it was through her that I got to know many people in Marseilles. I went with her to Naples on holiday.

I also made friends with Monsieur Amourousse-Bodin, who used to be *commissaire de police* in Paris, in the third *arrondissement*. When I knew him he had retired, and was farming near Avignon. He had a daughter named Christiane. At that time coffee was 5,000 francs a kilo. Every time my ship came back to Marseilles, I would bring him coffee, seven or eight hundred kilos at a time; the customs officials and police all knew me, so I used to help people out. I became friends with Monsieur Amourousse, and with his daughter Christiane. I started going to visit them whenever I was on leave.

That was where I learned about irrigation. I started with some land they were levelling for wheat. I would go out into the fields with Christiane; she had instruments with her for measuring the level of the land. She would hand them to me, and say: 'So, which is the higher ground?' and I would tell her what I thought. Then we worked on some land where they were going to grow grapes, putting in canals. After that, I really understood the work well.

I once worked with them for six months at a stretch. Later on, whenever I spent a month or two working with them, they would pay me 100,000 francs a month, which was a lot in those days. In the end, they wouldn't pay me by the month. I would spend my leave there, and go back to my ship when my leave was over. If I'd spent a month or two there, Christiane herself would open the safe and give me three or four hundred thousand francs. By then, we were engaged. So I was involved with irrigation, and with trade union activities; but all that time, I was working on shipboard. Once I took seven months' unpaid leave, and went to look at the countryside around Avignon, Bordeaux, Paris, Lille; just to see the fields of wheat and maize, the vegetables, the fruit trees.

In India I saw fields of sorghum, and oxen; fields everywhere, green all along the road. In 1950 I was working on a big cargo ship of the Compagnie Worms, travelling to India and New Caledonia. One day in India we went ashore. It happened a woman had died in the town that day. We had heard that in India they burn their dead, but we didn't believe it.

We saw a big funeral gathering, with lots of people. I said: 'Let's go over there.' Bullayi Malaado said: 'No, let's not go. If it were a Muslim burial, it would be all right. But they may not like us being there.' I said: 'Come on, let's go.' So we went. We found they'd laid the corpse out on a sheet of iron; they were murmuring words as they piled up wood on all sides, then above, until they were done, to make a fire that would burn its bones.

The others, Bullayi Malaado and Sili Jeremuna, moved away. But I moved nearer. I could see all the people. I took off my hat and put it under my arm. When they'd finished piling up the wood, their *moodini* gathered together and said a few words, I don't know what. When they had done, one person began to sprinkle petrol. He sprinkled it all over. Then he lit a match, and it caught fire. When I saw that, I covered my eyes with my hand, and left.

I said to Bullayi, 'Let's go that way.' There was some running water; we went to look at it. I was in front. We came to a large canal, irrigating large plots of land. I said to the man who was there: 'Why are you filling those fields with water like that?' He was Indian, but he understood some French; he even told me that he'd been to France.

He said these were all fields for planting sorghum. They irrigated the fields and let all the water soak in; then they tilled the fields, and sowed them. That meant irrigating twice, didn't it? I said: 'Yes.' He said the first irrigation was to wet the soil completely, so it would be moist deep down. The second was for sowing the seed. When the sorghum came up, they irrigated again, and added fertilizer; then they left it. Even in hot weather, the moisture deep in the soil rose at night, and gave strength to the plants so they didn't suffer. They left it like that until the plants flowered. Then they irrigated once more; and that was all.

He said, you could say they irrigated four times. But for them, it was more like two and a half times. Because their fields were very well laid out; when they irrigated, the water quickly spread all over, and as soon as it reached the edges of a field, they stopped irrigating that field.

I really saw irrigation there. We walked along the canals; those were real canals. When they irrigated, it wasn't like our perimeters, where you irrigate one field at a time. There, when the main canal was full, all the secondary canals would be full as well; the gates would be opened one after the other to let the water in. In one hour they would irrigate more than we can in a day. That was what I saw in India.

As for Madagascar, it was like my father's house, I travelled there so many times: on the *Ville de Massanga*, the *Colonel Vaizey*, the *Ville de Tamatave*. I often used to go and visit the countryside.

I'd heard that sweet potatoes grew wild in the bush there; when I asked about it, people said it was true. I knew a young mechanic who had a motorbike; when I asked him, he said he'd take me to have a look. It was the dry season then, and you couldn't see any sweet potato plants above ground. But if you scratched the earth, you would see the tubers lying there; as soon as it rained a little, they would sprout. He showed me a place where people went to dig up quantities of sweet potatoes. There were so many, they could never collect them all.

In Madagascar, they would capture wild horses and wild cattle in the bush. I saw the cattle and horses myself; I never saw them being captured, but I saw the net that was used, a net made by the *tubab*. When the animals went down to the water to drink, they would set up the net some distance away, then set on the animals from behind. The animals would start running. They would run and run and run; when they were tired, the people would start shooting at them with a drug that put them to sleep, and the animals would begin to fall. Whichever way they ran, there would be people with guns, shooting at them. Then they would be hoisted onto a lorry, and taken to a place where there was food and water; people went to look at them there. Afterwards they were sold as meat. I saw that in Madagascar. I saw things because I asked questions all the time.

There was a big garden by a river, where they grew onions and maize and cabbage. When I went there, they were planting out the cabbages. They'd made furrows of earth, with manure, and irrigated them. They said they always planted cabbages out in the afternoon. I took some myself and planted them with my fingers, then pressed the earth down. The man said: 'That's right, that's the way to do it.'

I asked him why they made furrows. He said: 'It's because we irrigate. When you grow maize under irrigation, or vegetables, you put them on furrows. Our fields are well laid out; when we irrigate, we wait until the water reaches half-way up the furrows, then we shut it off. Then the moisture rises.' It was in Madagascar that I first saw this.

I stopped travelling to New Caledonia, Tahiti, Japan, China, South America, and started working on ships serving the West African coast: Dakar, Conakry, Abidjan, Cotonou, Lomé, Takoradi, Pointe-Noire. From Pointe-Noire, I went to Brazzaville. In all those places, I made new acquaintances, who would come to meet me when my ship arrived. In places like Douala, people who wanted to go to France would bring me their papers and the price of the journey, and I would book their passage.

At that time Sékou Touré was active as a trade-unionist in Conakry, but France was making trouble for him, and he needed to leave the country. A man I knew, Pierre, brought Sékou Touré to meet me. He said he wanted to travel to France with his wife. I said he was welcome. We booked their passage, and they sailed with us.

We became friends. We used to play *belote*; we really took it seriously. He was determined not to let anyone beat him; I was determined not to let anyone beat me. We even quarrelled about it. The way I saw it, he was my friend, but I was more tolerant than he was. In the end, we would say: 'All right, you won today, but I'll win tomorrow.'

When we arrived in Marseilles, he said: '*Yugo*.' I said: 'Yes?' He said: 'I'm going to Paris, but I've nowhere to live yet. I need to find a place to rent. I'd like to leave my wife with you in Marseilles for the time being.'

At that time I had a hotel, 44 rue Nationale. Sékou Touré's wife stayed there for a while, in Room 2. One day he rang up; I was out, but he left a message saying he'd found somewhere to live, and asking me to ring back so he could arrange for his wife to travel to Paris that evening. I rang him and told him that she couldn't take the train that evening, but I knew someone who was going to Paris the next day; she could travel with that person, and arrive before nightfall.

When Sékou Touré went back to Guinea, I was still working on shipboard; I was still working on shipboard when Independence came. A friend of mine, Tall Beydi, ended up as *patron* of the Bar Ambiance in Abidjan. Sékou Touré wrote to him to ask, 'Is Jaabe So still alive?' Tall told him yes. Sékou Touré wrote back to say: 'Tell Jaabe I'm expecting him; I'm not a trade-unionist any more.' He put a paper in the letter for me to show at Conakry airport, so I would be taken direct to his palace. But I didn't go. When I heard about it, I said: 'That's good of him, but I don't think I can find the time to go to Guinea.'

The day Senegal became independent, I was on board the *Canada*, bound for Marseilles. A *tubab* said something to me that offended me at the time. He was a baker. He said: 'So, are you pleased now that you have Independence?' I said yes. He said: 'If only you knew.' I said: 'What do you mean, if only we knew?' He said: 'There's just one President in Dakar now, with authority over all of you. Now that you've become independent, you should have gone back to the way you were before.' I said: 'What do you mean, the way we were before?' In those days, I hadn't woken up yet; I thought like our first leaders. He said: 'The people of Fuuta should take care of their land, their roads, their taxes, their river. You should take care of your land, your roads, your taxes, your river. Then you could all join together later on. But the way you've gone about it, you'll soon be dominating each other, and no good will come of it at all.'

That's what has come to pass. At the time, I was offended by what he said; but I understood it later.

At the time of the Algerian war, the French expelled Arab workers; they needed other workers for their factories. I myself was in Marseilles then; I saw how people from the big factories in Paris would drive down to Marseilles and ask if any of us were looking for work. When we said no, they said we should write to our relatives at home and tell them there was work in France. That was what started people going to work in Paris.

When Senegal and Mali formed an association, I was in Dakar on leave. It was just after my daughter Bonko was born; I heard about it the day I went to get her birth certificate. At that time, I had already heard of Mamadou Dia.

Congo

Adrian Adams

The 'year before the call-up'. Isa Budu farmed with the other *taalibo* on Fode Mammadu's field at Papata. The year after, he left to join other members of his family in the Belgian Congo. About thirty-five men from Kuŋani were in Belgian or French Congo: mostly Tanjigoora, including a son of Sanba Haawa, with a few other *moodini* and members of the Jimmera family. The head of their community was Al-Haji MaJaaxon Ba, of the *Bundunko* household.

Al-Haji MaJaaxon Ba

N faaban ya ni, it was my father Ba Bintu Jaaxo, who started trading in Congo-Kinshasa. That was in 1911. He fought in French Congo; he went to Cameroun, and even settled there as an *alimaami*, but they expelled all the *alimaami*. He went to Gabon; he went to take Brazza's country; he went to take the country of Bangui, Centrafrique. Then he went to take the country of Fort-Lamy, Chad. Their general's name was Brazza. My father himself told me all this. He was there as a soldier; that was how he first went to French Congo.

My father went to buy cloth in Lagos, and took it to Congo-Kinshasa. After that first trip to Belgian Congo, he came back with a bit of money; in those days it was all coins, there were no banknotes. When he arrived in Xaayi, he bought two horses, fifteen head of cattle, and seven donkeys; he loaded three of the donkeys with kola nuts, and the four others with cloth. When he arrived here, he gave each household head a robe and a pair of trousers. *Ken da Kongo danŋen tinmi*; that was what beat the drum for Congo. After his first return here, other prominent townsmen also went there.

I myself arrived in Congo in 1930. I was already married, and had children here.

Isa Budu

O faabani saage telle gunne, our fathers began going abroad, then our brothers; they went to trade, and brought back money. They brought back *tubab* cloth, made up into clothes; they gave some to the family, and kept some for themselves. They used some of their money to buy cattle, donkeys, sheep. They bought meat to eat. In those days, it rained! We knew only sorghum then; we didn't know rice.

Our fathers left; we ourselves left. We went to Brazzaville and Lagos. We went to Niamey. All in order to earn some money. Sometimes you succeeded, sometimes not. If God willed it, you earned something. Otherwise, you could always come home. You could farm for your wives and children; you could harvest grain and store it in your granary.

Jaabe So

In 1961 I took leave; I went from Marseilles to Dakar, then to the Belgian Congo. I landed in Pointe-Noire, and took the train to Brazzaville. I arrived there at a time when you weren't allowed to go from Brazzaville to Léopoldville. Belgian Congo was dangerous then; it wasn't a good time for a visit. When I arrived at the house of relatives who lived in Brazzaville, I told them I wanted to cross over to Léopoldville. They said: 'You can't do that. No one goes to Léopoldville now; it's forbidden.'

I spent a month in Brazzaville; then I secretly crossed over to Léopoldville, by boat. I asked a Congolese I'd met, if he could find someone to take me across in a canoe. He said: 'It'll be difficult.' I said: 'All the same, if I go back to Senegal without visiting Belgian Congo, I won't have seen anything; I'll have wasted my fare.' He said: 'All right; I'll take you across for 5,000 Belgian francs.' I didn't tell the people I was staying with, because they'd have tried to stop me. I asked a friend of mine to tell them the next day that I'd gone to Léopoldville.

We set out at eight in the evening; we spent all night on the river, crossing towards the Léopoldville airfield, away from the harbour and into the bush. Anyone who's been to Congo knows that the river at Léopoldville is a very bad river; there are all kinds of dangerous creatures in it, and the current is very fast. At one point, I really thought we wouldn't make it.

We rested for a bit in the morning. I hadn't thought we'd be on the water for so long; I hadn't brought anything to eat. There was plenty of water, but no food. We met a fisherman in a boat, and bought some fish from him. There was a stove in the boat; we lit a fire, grilled the fish, ate, and went on. All this time, we were still out on the river. At five o'clock we landed near the Léopoldville airfield. I took my suitcase and said: 'Right, which way is the road?' There was a petrol station up ahead. He said: 'Be careful! If those people see you, they'll stop you.'

I was in a bit of a fix; I didn't even know which way to go. At five-thirty, I hid my suitcase and went to drink a *limonade* in the bar. They started calling me '*Ba-Sénégalais*', talking a lot, trying to make trouble. I was about to lose my temper, then I thought that would be foolish. If I did, they'd have me cornered; I didn't even know the way, and I couldn't very well ask them, 'Which way to Léopoldville?'

There were two men and a woman in the bar; I bought them drinks, one round, two rounds, three rounds. They became quite happy, and I made them completely drunk. When I saw that the woman was asleep on her chair, and the two men were starting to quarrel among themselves, I went out. I followed a path and came to the road. I stood by the side of the road; I heard a vehicle. A lorry came along; I held up my hand, and it stopped.

The driver spoke good French. He said: 'Where are you going?' I said: 'I'm going into town.' He said: 'You're Senegalese, aren't you?' I said yes. He said: 'I'll take you into town for a thousand francs.' I said: 'All right, just let me fetch my suitcase.'

He said: 'All my friends in town are Senegalese. There's one old man who's like a father to me.' I said: 'All right, take me there.' I got in, but he said: 'Don't sit there! Lie down, so no one can see you. I use this lorry to transport gravel; the *gendarmes* and police mustn't catch sight of you. Lie down; I won't stop for anything.'

I lay down; he took me to the door of a Senegalese. I didn't know him, and he didn't know me; he was a man from Tafasirga, called Mustafa. He'd been in Léopoldville for almost thirty years. We greeted each other, and he asked where I was from. I told him: 'Kuŋani.' He said: 'Ah! Like MaJaaxon Ba.' He told me: 'We'd better take your suitcase and hide it somewhere else. That Congolese may tell someone that he brought a Senegalese here.' They kept me hidden for two days. Afterwards, I wanted to go to MaJaaxon Ba's house; but the man's wife refused. She was a Bacili, Waraxiya Bacili. She said: 'So, you don't know me, but I know your house; it's as if it were my own. Chance brought you here; you must stay here.' So I stayed.

Afterwards, I said: 'I want to go up-country.' They said: 'No, don't go; it's too dangerous.' I didn't heed them; I took a plane to Luluabourg. I spent three days in Luluabourg; I went on to Tchicapa, then returned to Luluabourg.

When I returned there, the Congolese were fighting among themselves. The people of Luluabourg and those of Bakwanga, the Luluwa and the Balubakat, were fighting. You would see them in the town, with their *coupe-coupe*. The only other people there were Soninke; no administration, no *tubab*, just us Senegalese.

We were all in a bar, when we heard people shout: 'They're coming!' They invaded the bar. They took hold of the *patron* of the bar and threw him to the ground. I was sitting at a table with another Soninke, drinking *limonade*; I saw them take a *coupe-coupe* and cut off the man's head. I stood up and said: 'What's going on?' The man I was with took hold of me, and said: 'Look, So, take them as you find them; sit down.' They cut his head right off. The man got up, headless, and started running; he fell down in the street. They killed two or three men like that. I was beside myself . . . I said: 'They shouldn't do that sort of thing!' The man I was with said: 'Look, So, take a plane tomorrow, go home.'

We went back to where we were staying. More than eighty people were killed in the town that day; their heads cut off. It made me angry.

Afterwards, I wanted to go to Bakwanga. It was difficult to go there. I found a *camionnette* to rent, and reserved a place for myself; but another Soninke, a diamond smuggler, paid the man extra and took my seat. The car left without me.

Between Luluabourg and Bakwanga, they happened on some soldiers, who stopped the car. They took out all the luggage and searched it. There were forty-eight men in the van. They searched, and counted the money; it came to nine million. The Congolese said that if they just took the money, the owners would try and bring them to trial; so they decided to kill them. They killed them all, except for one man who was only wounded; he was in the bush for five days before making it back to Luluabourg. He was taken to hospital; he knew the names of the men who were killed. A patrol found their bodies, and the soldiers were arrested.

I was stubborn; I still wanted to see Bakwanga. While I was in Luluabourg, I met an army *commandant*; he took me there one day. I visited Bakwanga; there were many Senegalese there, and many diamonds.

I went back to Luluabourg and took the plane to Léopoldville. By then, you could travel freely. I spent the night in Léopoldville, and went to Brazzaville the next day, not secretly this time. When I arrived in Brazzaville, I found that my family in Dakar had written to say they'd had no news of me for six months. My relatives in Brazzaville said: 'You must go home.' I'd married a second wife the year before, who'd been promised to me from childhood: Faatuma Gaajigo, daughter of my uncle Denba Labbo Gaajigo with whom I grew up in Manayeli. I arrived back in Dakar a week after the birth of her first child, on the evening of the naming ceremony.

Coming Home

Adrian Adams

After Independence, Senegal continued to receive grants, loans, and technical assistance; spent mostly on public works, roads, harbour, buildings, to provide the appearance of a modern State, and on expanding the administration inherited from the French.

Mamadou Dia attempted to diversify agriculture and promote farmers' participation through *animation rurale* and the co-operative movement. After his imprisonment in 1962, Senghor's government concentrated on maintaining groundnuts as a cash crop for export. A French technical assistance company, the Société d'Assistance Technique et de Coopération (SATEC), promised to bring about a large rise in production to compensate for the large decrease in price when France joined the Common Market. Co-operatives in the groundnut areas functioned as instruments of State control, backed by large State corporations: the Office National de Commercialisation de l'Arachide (ONCAD), created in 1966, and the Société de Développement et de Vulgarisation Agricole (SODEVA), created in 1968.

This effort was not successful. Groundnut production fell sharply after 1965, because of drought and exhausted soils, but also because of farmers' increasing rejection of a system which drew them ever deeper into debt for seed, fertilizer, and ploughs, while profits from increases in the world price of groundnuts were kept by the State. The Common Market came to the government's rescue with a loan of two thousand million CFA francs, enabling it to forgive farmers' debts and stave off revolt. (CFA stands for Communauté Financière Africaine; throughout the period dealt with here, up to the 1994 devaluation, 1 French franc = 50 CFA francs.)

Industry, established before Independence to serve the whole of French West Africa, declined despite efforts to attract foreign capital. As the urban population increased, nearly doubling between 1959 and 1969, urban unemployment more than doubled.

Jaabe So

When I retired from the merchant marine, I spent the first three months in Dakar. When I left Dakar, I told my family, my wives and children, that I was coming here for a visit. I came here; no one knew what I was thinking.

I had got to know some people in Lagos. They had urged me to give up working on ships, and open a shop in Dakar; they would supply goods on favourable terms, because I'd done them favours in the past. But I decided against it.

I thought about what lies behind the way we live, and I saw that I'd been lucky. If I'd married and settled in France, I might have had a good car and lots of money; but to my way of thinking, I would have died without achieving anything. I would have lost my dignity as a man.

If you settle abroad, you can't say anything about your own country. You can't do anything about your own country. That was why I decided against settling in France or Dakar. If I'd stayed in Dakar, I'd have lived quietly in my household and grown big-bellied on meat and rice, unable to see things clearly, or speak, or do anything. I came back here so that I could do something.

Abroad and Home (1937–1968)

After what I'd seen in countries like Madagascar and South America, I wanted to develop my own country. If I'd stayed in France, I wouldn't have done that. My dignity as a man would have been lost; anything I'd done would have lost. I don't regret my decision to come home, even though it has meant many problems for me. It's better this way.

The year I returned to Kuŋani, I was at the Bakel *préfecture* when I heard that they'd arrested Mamadou Dia. When the *préfet* told me about it, I said Mamadou Dia shouldn't have been arrested; Africa needed men like him. I was just making conversation; but I wasn't pleased at the news.

The next day I was summoned to the *gendarmerie*; they asked me why I'd said that. I said: 'Because that's what I think.' They said: 'If the President arrests someone, you as a citizen should respect his decision.' I said: 'Do you mean that if the President arrests someone, a citizen isn't allowed to say what he thinks; he has to approve what the President's done? From what I know of Mamadou Dia, he's been good for our country. The President doesn't like him, he's arrested him; according to you, that means we have to say the President's right, even if he's not. Well, I won't say so. If you want to arrest me for saying Mamadou Dia shouldn't have been arrested, go ahead.'

So they arrested me. Afterwards they went and talked among themselves; then they came back and let me go.

That business about Mamadou Dia wasn't about politics. I don't care about politics. All my life, I've never been involved in politics. But I care about human dignity. I've travelled a lot, I've seen countries that are developed and countries that aren't developed. Mamadou Dia was good for us, because he saw that certain kinds of behaviour need to be done away with if a country is to make progress.

When I came here in 1963, I built rooms in the compound for my wives; they would come here each in turn, to take care of my mother who was growing old. One afternoon, when I'd finished the building and put doors in, I asked my mother to come, and said: 'I've finished. I'm going now. My wives will take turns to come and keep you company, they'll care for you and wash your clothes.' But she told me that wasn't what she wanted.

She said: 'Jaabe.' I said: 'Yes?' She said: 'Your father died a respected man. He didn't go to France; he didn't go to Congo. We had a good

Article 1: Shall constitute the National Domain all land not classified as belonging to the public domain, not registered, and whose ownership is not recorded at the Conservation des hypothèques at the date at which the present law comes into effect.

Article 2: The State holds land of the National Domain in order to ensure its rational use and improvement, in accordance with development plans and aménagement programmes.

Article 15: Persons who themselves are occupying and making use of land belonging to the National Domain at the date at which the present law comes into effect, may continue to occupy and make use of it.

However, the competent instance of the communauté rurale may disaffect this land, either because full use is not being made of its potential, or because the person is no longer using the land himself, or because to do so is in the general interest.

(Law No. 64–46 pertaining to the National Domain, 17 June 1964)

---

life, thanks to his farming. We never went hungry. We dressed well. We wore gold ear-rings. Now you say you're going away and your wives will come and care for me. So I'm to be lonely in life and the hereafter.' She pulled out the tucked-in corner of her cloth, where women keep money. She said: 'Is this what you mean: staying abroad and just sending me money? That's not caring. Caring is settling here.'

I said: 'Mother, I do intend to settle here.' She said: 'If you mean just bringing your wives here, and not coming yourself, then don't bother. I will go to my brother's house. They say that if a woman honours her husband, her children will never fail her; but that saying has not proved true for me.' I wept. That was when I made my decision.

I went to clear a field at Fanqanne. The river has taken away all the land my father farmed. It's all at the bottom of the river. When I was young, there was still a bit of it left for us to farm; but now the river's taken it all. Before, you would cross seven dips in the land before you reached the river; there's not one left now; that's how we know that the river has taken all our land at Fanqanne. The Geyi, too, had mostly land lying nearest the river; it's all gone now. I heard this, and saw it for myself.

When I'd cleared my field, I went to see Al-Haji Haamidu. I said: '*Moodi.*' He said: 'Yes?' I said: 'My mother doesn't want me to leave. She says I should settle here; God will grant me what He wills. I've decided what she says is true. So I've cleared a field. Otherwise, if I come and settle here now, I won't have any grain. If I leave the field in your hands, will you help me?' He said: 'God willing, it shall be so.'

He called the head of the Soroma *taalibo*. They burnt the field and tilled it. When it rained, they sowed it, and he sent someone to sow it again wherever the first-sown seed hadn't sprouted. When the time came for weeding, they weeded it. When I came back, the maize was almost ready to harvest. I went to greet the *moodi*; he sent for the head of the *taalibo*, and said: 'Show Jaabe his field.'

The evening I arrived in Kuŋani with my family, I went to my mother and said: 'Mother, I've come home.' She bade me welcome. She said: 'Surely God will grant you good fortune. Alone in this big house, I could not sleep for thinking, "Where are the people?"'

One day after I'd returned to Kuŋani, I saddled my horse and rode over to Tiyaabu. Madiwuri Siliman Bacili was the elder of Tiyaabu then. I told the people of Jonga: '*N ri biraadin muuri ya*. I've come for my living.' They went to tell the people of Tunkankaani.

Madiwuri Siliman told me through a praise-singer from Gande: 'The owner has come for what is his. According to the custom between us, we are to provide your evening meal, your midday meal, your morning meal and your clothes. You are *Manga-sakke*, the seed of Mammadu Denba from Kuŋani. The receiver of the gift has come for what is his according to custom.'

He said: 'Jaabe!' I said: 'Yes?' He said: 'Go and tell the *komo* to make an enclosure for the grain: let them take *melle* and bind them together.' They decided to make the enclosure at Konko Goola's house. They sent the town crier through the streets, to announce that *Manga-sakke* had come for his living; they said each *kagume* should give a *kande* of grain. It all came to ninety-seven *kande* of *gajaaba* sorghum, nine *kande* of maize, thirty-two *kande* of *ñobugu*. When they'd collected the grain, they sent for me. I rode over to Tiyaabu. They said: 'Here is your grain. How shall we send it to you?' I said I didn't know. They told the *komo* to bring a boat.

The Bacili women of Tiyaabu all came out. They said: 'This is the men's share. What about our share? Shall we make it groundnuts?'

Since the day I was born, I'd never asked for anything. My grandfather was born to that custom, as was my father; I went to make it new.

They filled two boats with grain and poled them upstream to Kuŋani. I said: 'I came to test the custom, to see whether it had grown old.' Madiwuri Siliman said: 'It is a cloth that cannot grow old. It cannot grow old in Tiyaabu. How could it? It was *Manga-sakke*'s from the beginning. If your cloth were to grow old in Tiyaabu, our rule would be at an end. In Tiyaabu there is only good for you.' He said that at the time of *laxaasara* prayers; all Tiyaabu was there.

People in Dakar said: 'So, you won't be able to live in Kuŋani after twenty years in France.' I said: 'Yes, I will.' They said: 'What are you going to do—open a shop?' I said: 'I know what I want to do.' Because I intended to farm.

The field Al-Haji Haamidu had farmed for me, yielded one ton two hundred of grain. At that time, maize grew well at Fanqanne. From then on, I farmed for myself.

My father used to let people use his land freely. The land that he himself used to farm was right by the river, and the river had taken much of it away. At that time, Mahamme Kamara and his family were using the land once farmed by the runaway slave Ñaxana. I was ashamed to say it, but in the end I said to Mahamme: 'Look, I need that land to feed my family.' He still has the land further upstream, that my father saved for Kebe Ñaxana.

That dry season, I crossed the river and went to stay with Mahamme Moodi in Sélibaby. We went to look at a shop that sold farming equipment. The owner was a *tubab* named Roger. We recognized each other at once; we had known each other on board the *Jules Verne* in 1942, before he left to serve on a submarine destroyer called the *Javelot*. He asked me what I wanted, and I said: 'Ploughs.' He showed them to me, and said he could let me have them on credit, as Kuŋani was near Bakel, and he often went to Bakel. They cost 3,000 francs then; I would pay a thousand a year.

He gave me four ploughs, and I brought them here. I kept one for myself; I gave one to Al-Haji Haruna Tanjigoora in Golomi, one to Mahamme Kamara, and one to Mahamme's brother Samaane in Bakel. They were the first ploughs ever in the Bakel area. The first year, no one used them; they didn't know how.

I myself chose an ox and began to train him. People laughed at me. I heard them saying: 'Why's he doing that? Says he's going to farm with an ox? Nonsense. Just use a hoe, that's all.' With God's help I trained the ox and ploughed with him at Fanqanne. The harvest was good.

That year the river flooded. There was a big boat in Kuŋani called the *Ancerville*, named after a passenger ship that did the Marseilles–Pointe-Noire run; the boat went back and forth to fetch the unthreshed grain. It filled my largest store-room.

I gave my oxen ships' names. The first pair of oxen I trained were called *Hoggar* and *Ville d'Amiens*. The second pair were called *Mangin* and *Canada*. I helped Mahamme Kamara buy a horse, and he ploughed with it the next farming season. Those were the first ploughs to be used here. People criticized me; but when I was ploughing at Fanqanne, they would leave their fields to come and watch.

In 1963 I went and asked for my grandfather Waali Silaamaxa Sisoxo's land at Seegankaani. I already had land in Kuŋani, but I had the idea that someday I might like to try irrigated farming.

That land is called *dinaare* land. It was given that name long ago, because it is such good land. You can't farm it and not have a good harvest. Any crop you plant there will succeed. The *dinaar* is an Arab coin from the days of the Prophet's disciples.

The land at Seegankaani had lain fallow for fifty years, when the Kuŋani *debegume* Salun Geyi went to see my grandfather Waali Silaamaxa in Bakel, and asked him to let him have the land to share out among his people; they would pay *jakka*. My wife Maamu's elder brother Famaxa Sisoxo said my mother and I were entitled to the land; I should take as much as I needed, and let other people use the rest, on condition they paid *jakka*. The other people who farm at Seegankani, send a token *jakka* to Bakel. Even if they harvest two tons, they just send four or five *muudu*, to show that the land is not theirs. But I don't pay *jakka*, because my land at Seegankaani was given to me outright.

I spent 35,000 francs to have the land cleared: just for cutting down trees. One day when we were at work clearing the land at Seegankaani, I came across a lion lying under a *xiile* tree. It had killed a warthog and was eating it, with its whole head thrust inside the carcass; when it lifted its head, it was covered with blood. I had to chase the dogs away so they wouldn't catch sight of it. When I went back later, it had eaten half the warthog and was lying asleep. If I'd had a gun with me, I could have gone up and shot it.

At that time, there were still lions, wildcats, warthogs, and monkeys to be seen by the river. But the *tumujo*, the smaller animals, were starting to disappear. And you had to go as far inland as Jaabalu or Guñan Yaamadu, to see a deer or an antelope. There were still fish in the river, and in the creeks and ponds. But then the fish began to disappear.

When the land had been cleared, we set fire to it. The fire burnt so strongly that people in Bakel, Kuŋani, and Bema all said that the bush was on fire. After that, there was still the *bodo* grass to uproot. I paid twenty-five thousand for that to be done.

The first year I farmed Seegankaani, in 1964, I grew maize and *gajaba* sorghum. The yield was so great that even with the help of all the women in my household, I couldn't get the harvest in. I finally asked the town's *komo* to help me. I had so much grain that there was one full granary still left at the end of 1965; I threshed and sold it.

The second year I farmed Seegankaani, I cleared and burned Fanqanne, and gave it to Isa Faatuma Silla Tanjigoora to farm. But I said: 'If it's flooded, I'll farm it myself!' He agreed. I ploughed all of Seegankaani with my oxen. The crop had grown tall when the river flooded and drowned it all. The flood also drowned the fields at Fanqanne. I sowed Seegankaani again with *gajaba* and maize, and Fanqanne with maize.

All Kuŋani knows that I harvested so much grain that year, I didn't know where to put it. We don't usually thresh sorghum in the field, but the *komo* went to thresh the *gajaba* and winnow it and put it into sacks, then brought the sacks to town in a boat. I harvested four tons of *gajaba*. Isa Sabeli measured it all with a *muudu*; thirty *muudu* equals one hundred kilos. At Seegankaani, I also harvested two boatloads of maize. After that year, I was keener than ever to farm there.

I harvested two *Ancerville*-loads of maize at Fanqanne as well. At that time there was a large boat here, named *Ancerville* after an ocean liner that did the coast of West Africa; it could hold eighty *kande* of grain at a time, enough to fill a granary.

In 1963, 1964, 1965 the rains came and there were good harvests everywhere, by the river and inland. It was from 1966 on that things began to go wrong, and yields grew smaller and smaller every year. Crops planted on low-lying ground might do all right, but the same crop on higher ground would die before bearing grain.

That made me start thinking about irrigation; because I'd seen irrigation before. I kept thinking how I would go about irrigating my field. In the end I decided to go and work in France, in order to bring back a pump.

Adrian Adams

The 1960s were years of drought for River people. More and more men left to work in France. In 1964 an agreement between France and Senegal restricted entry to immigrants who already had a contract with an employer. African immigration to France continued to exist, now more often than not illegal.

In 1964 the Senegalese government adopted a *Loi sur le Domaine National* placing all land not registered as private property, i.e. 97 per cent of Senegalese land, under State control. Land tenure as known to the River people continued to exist, but unsanctioned by law.

In 1965 the Société d'Aménagement et d'Exploitation des terres du Delta du fleuve Sénégal (SAED), a State corporation, revived large-scale irrigated rice-farming in the Senegal River Delta. Thirty thousand hectares were to be put under rice cultivation in ten years. Six new villages were built between 1964 and 1967; about 20,000 people were brought to settle there. But there were difficulties. In 1968 20,000 acres yielded only 500 tons.

Jaabe So

On 16 April 1968, I left Kuŋani to go to Bakel. At that time, it was nearly ten years since I'd left France to come home; but that day when I went to Bakel, I had France on my mind. I'd told my family I was going to Bakel; but in Bakel, I took a car to Dakar.

Soon after I arrived in Dakar, the *Général Mangin* came up from Pointe-Noire. Sanba Sumaare was on board, working as a stoker. He came to the house where I was staying, and told them to tell me that the *commissaire* wanted to see me straight away.

I went on board. The *commissaire* asked me how I could disappear so completely, after working on ships for so many years. I told him I'd wanted to work out a way of settling down, but I'd probably be going on to France sometime soon. He said: 'No, if you're going, fetch your things and go with us.' So five days afterwards, I sent my family a telegram saying I'd arrived in Marseilles. I also sent them 100,000 CFA francs.

In Marseilles, I went to the Compagnie Fabre. They said I should wait for a ship to return to harbour; as a *titulaire*, I was certain to be taken on. Then I went to the Compagnie Worms. They said that at the end of the month I could join the *Ville de Massanga*, bound for Saigon, as a baker. I didn't want to go to Saigon; so I just said 'Yes', and left. I went to Messageries Maritimes; they said the *Ville d'Amiens* would be going to New Caledonia, but only in two months' time.

I went about visiting people. And I thought: 'It's funny; I left here thinking I'd go and make plans for taking up farming in retirement, now I'm back here with a different plan, to buy a pump because of the drought.'

Madame Claire was dead; Monsieur Amourousse was dead; so were my other friends, like Madame Dubois or Monsieur and Madame Amblan. I felt I didn't want to stay in Marseilles, and decided to go to Paris. When I said I was going to Paris, my friends Papa Jallo, Dawuda Jabira, and Banjugu Jawara were all angry with me. Papa Jallo said: 'You won't be able to stand it there.' He and Banjugu were so angry they refused to accompany me to the railway station.

PART II

Paper

4

Paris
(1968–1973)

Adrian Adams

In 1968, the year Jaabe So went back to France, Senegal exported mainly groundnuts, and imported foodstuffs, consumer goods, fuel, and equipment, almost exclusively for urban use. Deriving almost half its revenue from customs duties on imports, the government sought to maintain exports, rather than reduce imports. The Dakar International Fair was created to attract capital investment. Investment in tourism was welcomed. The US-based corporation which set up BUD-Senegal, a large export-oriented market-gardening scheme, received major tax concessions; as did the French corporation which set up the Compagnie Sucrière Sénégalaise (CSS) at Richard-Toll in the lower Senegal River Valley.

SAED's rice-growing schemes in the Delta cost five and a half thousand million CFA between 1964 and 1972. Farmers were provided with fertilizer, seed, and hired machinery on credit, and were closely supervised by SAED personnel. They worked only forty-odd days a year, sowing and threshing; SAED did the rest. SAED's services were expensive; also, given the lack of subsistence crops, farmers tended to consume much of their harvest themselves. These factors, combined with technical problems, drew farmers ever deeper into debt. The year's debts often exceeded the cash value of the year's produce, and debts accumulated from one year to the next; at the end of the 1969/70 season, unpaid debts averaged 8,000 CFA francs an acre.

Like their counterparts in the groundnut-growing areas of Senegal, the Delta co-operatives were essentially administrative devices for marketing and credit allocation. SATEC was called in as early as 1967, but could do little. After 1968, farmers left the Delta in increasing numbers, and SAED suspended plans for bringing further surfaces under cultivation.

The drought intensified, and ever greater numbers of men left to seek work in France, even though conditions for African workers in France were steadily worsening. In 1972 the River Valley suffered total crop failure.

Jaabe So

N ga joofe Pari, when I arrived in Paris, I went to live in the 14th *arrondissement*; then I moved to the 13th *arrondissement*, to a *foyer* in the rue Léon-Maurice-Nordmann. I looked for work. I was offered work as a dustman, but I turned it down; I didn't feel I could do that. I was offered work at the Renault factory; but when I went there, and saw the conditions of black workers there, I refused. Then I was offered a job at the *Armée de l'Air*, looking after their keys. But I felt that was not a proper job, so I left.

I'd met an officer there who remembered me from the *Emile Bertin*, before the landing at Toulon, and knew I had a cook's professional certificate from 1942. He was a lieutenant at the time. He was retired now, and walked with a cane; but he took me to see the colonel in charge, and they asked me to prepare *couscous marocain* for them on Sundays, with thirty chickens and half a sheep. They paid me 80,000 francs a time.

During the week I worked in a factory in Saint-Denis. When the *patron* hired me and saw my papers, he said I wouldn't like the canteen; he had a gas cooker and fridge brought in, so I could cook for myself.

The *foyer* rue Léon-Maurice-Nordmann was very crowded. The *concierge* told me that the owner had planned to add an extra storey, but found he didn't have the money. I was treasurer for all the Soninke in the *foyer*; each town's *caisse* was in my keeping. I told the owner I'd advance him the money, without interest, so he could build an extra storey, fifteen rooms, and people would be less crowded. He signed a paper, which I kept, and I give him one million five hundred thousand francs. He brought bricks and cement, and started work straight away. I advanced him four million francs in all. When the rooms were ready, they were shared out among people from Kuŋani and Golomi, three to a room. They paid a month's rent in advance. The owner took in two million the first month, one million the second, one million the third; he paid back all the money he owed me. People only heard about it when the work was done.

Afterwards they started building government hostels. They did that in order to get hold of the money kept back from workers' pay. Eight people to one room, each paying 15,000 francs . . . If you don't have a room, you have to accept that.

All the time I was a seaman, I had French friends. There were 680 African seamen in France then, from Madagascar, Cameroun, Senegal, Mali; no more than twenty Malians. In Marseilles people greeted me, in the streets, everywhere; I greeted them, we talked, we invited each other, we knew each other. I had friends who were like family. Even in Paris, when I first went there in 1947, life was good.

The first France I knew was after the war. The second France began when they expelled the Arabs. Racism began then and has continued ever since. If you're an Arab born in France of an Arab father: get out! If you're a *tubab* born in France of an Italian father, or Spanish, or Portuguese, you're not a foreigner; it's Arabs and Blacks who are foreigners. They need to take a good look at all that. There are people in Marseilles now, local government officials, police commissioners, who were born in our day: all foreigners' children. They count as French now. If you're looking for real Frenchmen, where are they?

I came home, then I went back to France; to Paris, with the *émigrés*. I went back to France for the second time: the last time, that's what I hope, let it be my last time in France. The way people live all crowded together in the workers' hostels, the way they travel in the Métro, the way they have no personal relations with Europeans: I didn't like that.

I found that way of life very, very hard. No African can say he's respected in France: none, none, none. From time to time I would say in the *foyer*: 'How can you live here?' In the Métro, if you had words with a Frenchman, he would say: '*Fous le camp chez toi.*' I myself once had words with a woman in the Métro. She said: 'Why don't you just go back home; *qu'est-ce que vous avez à venir m'emmerder ici?*' I answered: 'Yes, you can say that today. Time was when we entered Paris, there were Germans in control of the Métro, and you were hiding in your beds, you wouldn't even come down into the street. People had to take you your drinking-water. In the meantime we were down here, spilling our blood on the ground. Now you can say that to black people.'

That's what really revolted me; all those remarks, everywhere. If Africans think they're well treated in France, it's not true. A black man in France is like a dog. Not even a good dog; a stray. A stray dog. The administration, the police, the *gendarmes*, women, young people: everyone in France has heard the news that black people are dogs.

I was really surprised to see France go wrong like that, so quickly. I would say in the *foyer*: 'We'd best forget about the money we earn here. If a Frenchman speaks nicely to a black man, he wants something from you. Otherwise, he doesn't care for you at all. There are French people in our country; they're made welcome, and people respect them. But here . . . Everyone here knows that when a white man kills a black man, there's no justice. No justice at all. God judges.'

If you don't have work at home, you'll endure anything to feed your family. It's better to take up farming again, to develop our land, than to work in France. I know France: I saw the good side of France, I've ended up seeing the bad side.

If you don't know, you think it's marvellous there. . . . If you have money and go to a hotel, you don't know France. It's the people who work in France, who know France. We never thought Africans would suffer in France. We thought that when you go to France from Senegal, you're still at home. It's not true. It's not true. What you feel when you're there—you can't explain it here.

Disrespect begins in the city streets; it extends to the cemeteries. Six months after an African dies, they dig him up and burn him. When we realized that, we set up a special fund so that people who die in France can be sent to be buried at home. Now, we know that's a waste of money; a person should be buried wherever he dies. But burial in France is not proper burial. There is no respect there for us, dead or alive.

The French who are in Senegal don't care about the bad things in France. They're here, telling us what to do. When I see them, I think of what they've done to us. All the good we did in France has been repaid with evil. I'm in my last life now; no more going to France for me. When my son grew up, I could have sent him to France. But I said no.

One day I had some money saved up. I saw a pump worth 400,000 francs. Later I went back to the shop and discussed the price; next day I went to pay for it. That's when I met Uccelli, a European who was interested in agriculture. When he saw me paying, he waited until I had finished. Then he asked me: 'Where are you from?' I told him: 'I'm from Senegal.' He said: 'That machine there—what do you plan to use it for?' I said: 'Farming.' He said: 'An African working in Paris who buys farm machinery deserves to be helped.' That's how we met.

In those days it wasn't called GRDR [Groupe de Recherches et de Réalisations pour le Développement Rural dans le Tiers Monde]. It was Uccelli, Dubois, De Coninck, Madeleine, and me, Jaabe So. I was the only African. When it was set up, they asked the *Ministère de la Coopération* for money, saying they were going to train *émigrés*. They were given fourteen million francs. They rented an office and hired a typist. Uccelli became president, and Madeleine, secretary.

One day I said in a meeting: 'That money was given in the name of Africans, but they don't know anything about it. How are we going to let them know?' Madeleine said: 'Monsieur So is right. Give us an idea.' So I said: 'Why not hire coaches on Saturdays and Sundays, to collect people in the *foyers* and take them to see farming in the countryside?' They agreed.

De Coninck's grandson once said to me: 'You should go in for *développement paysan*. You lot are farmers; the GRDR isn't for you. The money was given in your name; at least one member of the board ought to have been an African. But it's not for you.'

I left the GRDR; I stopped going to their meetings, and no one asked me why, except Dubois who wasn't happy that Uccelli had been made president. They followed their idea of development, and I followed mine.

Through De Coninck, I met a Madame Bodin. She wanted to hire me to work at her château, but I said I was going home. She had a nephew, Richard Elsner, who worked in London.

One day Uccelli said: 'So, do you want help; money, say?' I said: 'No, I'd rather a little development, if you can help me find a technician or meet someone at home who knows about technical things.' He said: 'I don't know Senegal, I've never worked there.'

Later on he said: 'I'm not rich enough myself, but I can help you find a financier, and we'll look for a technician. We'd best write to the Senegalese government though; what if they refuse?' 'Ah,' I said, 'that's something else altogether.' Uccelli wrote to the government, and the government wrote back to say 'Yes'. He came to see me that evening. He said: 'Good, they've agreed; now I'll look for a financier. But how are we going to find a technician? All the young technicians in France already have jobs. We'll start looking now.'

Well, there we were. Two months, three months—Uccelli still hadn't found anything. One fine day, I was talking about this problem

> *Those Senegalese students who proclaim themselves Maoists seem to have no intention of going to the villages to work with peasants, win their trust and take part in an uprising. If leaders were to emerge from the mass of the peasantry, the situation might evolve more rapidly; let us refrain from prophecies, if not from hopes.*
>
> (René Dumont, *Paysanneries aux abois*, 1972, 212)

> *The firm of Norbert Beyrard has presented an integrated development programme for the Senegal River basin which begins with large dams and ignores the potential of small-scale hydraulics. A familiar tune: large-scale projects bring profits for construction firms, banks, and consultancy firms.*
>
> (Ibid. 228)

> *Senegalese peasants have been driven to reject a form of 'progress', a system of modernization, which makes them ever more dependent, lowers their standard of living, and compromises their very dignity; they are now turning back towards a subsistence-oriented economy....*
>
> *Privileged Black urban minorities have in part taken the place of the White colonizers; through their plundering of public funds, their disregard for the common interest, and their alliance with neo-colonialism, they constitute for the most part a parasitic class, which therefore deserves to fall.*
>
> (Ibid. 229)

in the place where I worked. There was a storeman there who was always reading the papers; he said: 'Hey! Here's a chap looking for work; he's a technician.' He gave me the paper, and I sent it to Uccelli. He wrote to the technician, Robert Aprin, and they came to an agreement. Robert was hired by a company called CIDR [Compagnie Internationale pour le Développement Rural]; the money to pay him came from Richard Elsner in London. Uccelli ended up leaving GRDR for CIDR.

I went to tell Uccelli: 'I'm leaving next month.' He said: 'Go ahead. Not this year, but next year, you'll have your technician.'

Adrian Adams

Crops failed throughout most of Senegal in 1972. Shortly before the legislative and presidential elections of January 1973, a 'solidarity campaign' was launched in the pages of the government-controlled newspaper *Le Soleil*. In June 1973, 524 million CFA francs were distributed to peasant farmers. A household head who had lost his entire crop was entitled to 509 CFA francs per household member.

In 1973 SAED extended its operations to Dagana and Nianga in the Lower Valley. In 1974, an official decree made SAED responsible for the agricultural development of the Senegalese bank of the River, from Saint-Louis where its headquarters are, to the Mali border. In 1975 SAED moved further upstream to Matam in the Middle Valley, where an expatriate SATEC agent, originally sent to the area to improve sorghum production, had just set up three small irrigated rice-growing perimeters, twenty-five hectares in all, involving about one hundred and fifty drought-stricken farmers.

The Organisation pour la Mise en Valeur du Fleuve Sénégal (OMVS) was created in 1972; its programme was announced in 1973. It was centred on the construction of two dams, one upstream in Mali, the other in the Delta, which were to open the way for intensive irrigated farming. The ultimate goal was to bring three or four hundred thousand hectares under irrigation. The main crops would be rice and wheat.

In 1973 CIDR signed an agreement with the Government of Senegal, according to which it was to assist in the reinsertion of returned emigrants from the Bakel area. Jaabe So did not know about this; he was never told.

5

Hope Traduced
(1973–1975)

Year One: 1973/1974

Jaabe So

N d'in mesin ri. In 1973 I came back with my pump. People watched me build canals. I heard them say: 'What's he doing? He must be planning to put pipes in there.' I built canals and dikes. Then I planted my field. My pump was a little Italian model that ran on petrol. Mornings I would start it up at ten and stop it at twelve; afternoons, I would start it up at five and stop it at six-thirty. My eldest son Mammadu was fifteen then. One day the pump broke down. I said: 'Leave it, tomorrow we'll take it to Bakel.' Later on, someone came and told me he'd gone to my field and found Mammadu taking the pump apart. I was very angry and went back to Seegankaani to tell him off. As I drew near, I could hear the pump working. He said he'd found a bone stuck in the motor. I didn't say anything, but he put it all together again.

I grew maize, and harvested it all. I had to ask the *komo* to help me with the harvest. The tithe alone was ninety *kande*. For a household that pounds three *muudu* of grain a day, that's enough to live on for two years. That was what I harvested at Seegankaani. None of my crop was lost to drought; not a thing.

When people saw what irrigation could do, they all fell in love with it. In the end, people would come to me and say: 'Jaabe, I've done my field like yours; lend me your pump.'

I came to see that it was not right to keep this work for myself alone. I had another idea.

Koota yogo, one fine day at eight in the morning, Robert turned up in the car from Bakel; he asked for 'Jaabe So's house'. I was standing right there myself. Well, I brought him home, and made rooms ready for him and his wife.

The aim of the Integrated Senegal River Basin Development Programme is to provide the people of the Senegal River Valley with adequate basic nutrition and increasing revenue in cash, thus enabling them to move beyond the uncertain subsistence economy in which they live, into a modern consumer economy. . . . However, in order to ensure funding for the programme, the mechanisms established must allow for reinvestment of a substantial part of the profit. The crops chosen must, while meeting the subsistence needs of the population of the valley as quickly as possible, generate abundant cash flows making it possible quickly to reach the stage of economic take-off. Without going into the detail of the various types of land use which will be chosen by different States or irrigation schemes, they can be reduced to two main categories:

(1) Industrial plantations where peasant farmers are entitled only to a wage, the crop being the property of the plantation. For this type of concern, it is possible to reinvest the entire cash flow. It may be noted that from this point of view it makes no difference whether the plantations are managed by private companies or by State companies of socialist inspiration.

(2) Schemes governed by the different States' laws on national domain, where land remains the property of the State, but peasant farmers are entitled to farm the land and dispose freely of the crop, so long as they pay for the services provided by scheme management (inputs, water, construction work, technical advice) and pay a fee which makes it possible to repay loans and fund projects.

(OMVS, Programme intégré de développement du bassin du Sénégal, *1973*)

He said: 'Let's go and see your land. We'll farm a big field for you there; people will come and look at it, and anyone who wants to learn can work in your field.' I said: 'No. First let me introduce you to the people here.'

A meeting was called; all Kuŋani was there. I said: 'Here's a technician for us. I asked a friend to send us one, to help us with our work. Here he is.'

Robert said: 'What are we going to do?' I said: 'Listen, the best thing to do is to make development for everyone who wants to take part.'

I held a second meeting in Kuŋani. I said: 'Let everyone come, so that we can speak together. You elders, you have travelled and lived. I wish you would help me make development, you who know what it is. If you don't help me, that means that you want our country to remain undeveloped.' The elders said: 'Very well, we'll help you until three years have passed.'

Robert said he hadn't any money left; his salary hadn't yet been paid into his account in Dakar. I gave him 40,000 francs, so he could settle in and go back to Dakar after his money. All the time he was in Kuŋani, I provided for him and his wife.

Al-Haji Mpali Tanjigoora

O ku jamaane, it is farming that makes our country good to live in. No matter how much you earn in Congo, the Ivory Coast, or France, unless you take up farming again on your return, your wealth will be worth nothing; it will all vanish into the cooking-pot. When the rains failed, we didn't know what to do.

When Jaabe came back, he gathered everyone and said we should farm together. When he explained what he had in mind, everyone wanted to take part. He said that if we wanted help, we should first help ourselves; for the first two or three years we should not farm each for himself, but all work one field together, to gain strength.

Jaabe So

Non ya ni, it was in Madagascar that I first heard of collective fields. I went to visit a place by a river where people were growing rice. A man showed me a large field he said was their *champ collectif*; they all worked there together, sold the crop, and put the money in their *caisse*. Their own personal fields were a bit further on.

There's a woman in Kuŋani named Xujeeji Dabo, a neighbour of mine; ever since we were young, I've called her Tuntunba, which means a fat person, to tease her. After that meeting, Tuntunba came to see me and said: 'Is that work for men alone? Women are farmers; women would like to take part as well.' I saw that she was right; and I encouraged her to talk to the other men about it. Some men were against the idea; but in the end, they agreed. So women became members of the Kuŋani group. When other groups were formed, I told all of them that women should be allowed to join; they all agreed. Everyone can see now that having women in the group is a good thing.

Modibo Koyita Tanjigoora

Rober ga ri, when Robert arrived here, he went to stay in Jaabe's house. The whole village gathered to meet him. He told us that he had come to help us farm. These past few years, there was not enough rain,

Hope Traduced (1973–1975)

and we'd not grown enough to live on. He wanted to help us so that we would always have enough to live on. He also said that he hadn't come to make us work for him, nor would we have to pay him; his pay was being sent from France.

The *debegume* chose four people to accompany Robert through the bush, to show him the places that are good for farming and tell him what grows best in each place. Since Robert came here through Jaabe and was staying in Jaabe's house, Jaabe was chosen first. I was chosen fourth. We set out on foot. As we were walking at Fanqanne, along the river bank towards Golomi, Robert pointed out several places he said would be good; when he heard one of them was Jaabe's, he said we might as well choose that one. When we got back we told the *debegume*; he called a meeting, and every person who paid tax gave 100 CFA francs towards fencing the field, so that we could make a vegetable garden there. We also had a well dug for 37,000 CFA.

Jaabe So

N faaban ñiiñen ya ni ke; that was the land my father used to farm. Then we went to other towns. We first went to Golomi, and I introduced Robert. Then we went to Arundu and Yafera. We held a meeting in Yafera; they didn't agree straight away, nor did Arundu and Baalu. Then Yafera and Arundu said 'Yes'. We went back to Baalu; they had no land for a vegetable garden, but they agreed to grow a big field of sorghum.

Mammadu Lamiina Bacili

Jaabe ri Arundu, Jaabe So came to Arundu and said he would like us to farm together. The third time he came, some people agreed; they said the work should be for old people and young, men and women. That was in 1974.

Jaabe So

A ña kundun ya, that is how it was. Once the villages upstream had all agreed, we came back to Kuŋani. Then I took Robert to Tiyaabu. The second time, I was busy with the mosque we were building, and sent him to Tiyaabu with Modibo and Hamme. When he came back, he said to me: 'If I go there without you, they won't talk to me. Don't let me go there on my own.'

I went with him to Tiyaabu, once, twice, three times. The people of Tiyaabu said: 'All right: we'll grow a field of sorghum, and a vegetable garden during the dry season.' We went to Yelingara. We went four times to Jawara, a large town. The fourth time, we went there in the morning; there was a big meeting. I said: 'Look, I brought this man here to help us, to show us how we can develop our country.' People said: 'What does that mean?' I said: 'You grow tomatoes and cabbages, you grow a big field of sorghum; you carry on like that, working together, until we manage to develop our country.'

Well, Jawara agreed to grow sorghum together the first year. In Mudeeri, they didn't agree the first year, although they did the year after. So I went on to Gallaade; they agreed straight away, and showed us the place where they wanted to put a vegetable garden. In Gande too they agreed straight away. In Gaabu and Guñan, in the *jeeri*, they said they would plant vegetable gardens.

Manayeli Jallo

Jaabe do ke golle ga ri, when Jaabe brought this new way of farming here, he didn't keep it for himself. He could have worked just for his household. But he didn't do that. He brought the work out in the open, for all Kuŋani to take part. Then he went to visit other towns. In every town he treated people alike. He didn't favour one set of people over another; he went to see the *debegume* and asked him to gather the townspeople.

Here in Manayeli, he told us that he had learned a way of farming that he wanted to show us, his relatives. We should unite and farm together, and secure our claim to our land. He said we should list the names of men and women separately. At the beginning, every man and woman in town joined in.

Mammadu Bakari Saaxo

Jaabe ga ri Jawara, when Jaabe came to Jawara, he spoke to the assembled elders. He said: 'I've brought a way of farming to show you, that will benefit all of us. Its name is *champ collectif*; it is for everyone.'

Denba Giise Saaxo

Jaabe So da golle ke riti, it was Jaabe So who brought this way of farming here, from Baalu to Gande. When he first came, few people understood what he was saying. For most people, 'I'm going to help you' means 'I'm going to put money in your pocket'. Whereas Jaabe said, 'I'm going to help you learn.'

Hope Traduced (1973–1975)

Jaabe So

O ga daga Golomi kootan be, the day we went to Golomi, my nephew Bokar Sisoxo from Bakel, an Inspector of Schools, and a schoolteacher friend of his from Tiyaabu came to spend the night here, to ask what Robert was planning to do. Robert said his idea had been to work with me alone, but I had a different idea; I wanted each town to form a farming group. Bokar said he thought my idea was good; that way, they'd be able to join in. When they went back to Bakel, they organized the farming group that became known as Bakel-Gassambilaqe, after the name of the land they chose to farm.

At that time, Papa Kane from Golomi, who was like a son because my wife Maamu looked after him while he was at school, was the director of ENEA [Ecole Nationale d'Economie Appliquée] in Dakar. He sent some ENEA students to spend six months in Kuŋani as part of their studies.

Year Two: 1974/1975

Jaabe So

Ken siine, n da Seegankaani soxo; I farmed Seegankaani that year. The Kuŋani *groupement* grew a collective field of sorghum and maize. There were 168 men in the group, and seventy-five women. The rains were good.

During the dry season, we dug a well on the land I gave them at Fanqanne, and grew vegetables: tomatoes, onions, cabbage, lettuce. We sold the vegetables and earned 100,000 CFA francs for the group's *caisse*.

We spent 125,000 CFA francs on preparing for the next rainy season. We bought four oxen for ploughing; I trained them myself. Kuŋani people working in Congo and France sent us money; the young people's association also gave some. The people in Tiyaabu, Manayeli, Jawara, and Baalu also grew sorghum; Tiyaabu, Manayeli, Gallaade, Gande, and Arundu also grew vegetables in the dry season.

Manayeli Jallo

O siine fana, the first year we farmed together, we grew sorghum. We all worked together every Thursday. After the harvest, we threshed the grain, winnowed it, and put it in sacks. When the price rose, we sold it and put the money in the group's *caisse*. In the dry season, we grew vegetables.

Denba Giise Saaxo

O su be ga kafi, all of us who joined the group farmed a large *champ collectif*. That was before we had a pump. We grew *feela* sorghum, *ñeniko*, maize, and okra. The grain we harvested earned us a lot of money. It was that which gave birth to our co-operative store, here in Jawara.

Al-Haji Denba Bacili

Ken waxati, n ñi golliñaana Faransi; I was working in France at the time. In 1974, when I came back to Arundu to marry, I found that they had planted a vegetable garden just below our household's new compound, with water fetched from the river. I helped water the garden. I was happy to see it.

Modibo Koyita Tanjigoora

Rober ti ñiiñen ke laato fanŋe. Robert said that the land at Fanqanne was good, but a bit far from the river. So we all went to look at the land downstream, towards Bakel, at Sanba-Salu. He said: 'It's good here.'

Al-Haji Mpali Tanjigoora

O daga Sanba-Salu, we went to Sanba-Salu with Robert. He started at the end near Al-Haji Modibo's field; he dug holes there, and looked at the soil. Then he went further downstream, where Jaabe's land was, and ours. He said the land was better there.

Mammadu Lamiina Bacili

Jaabe do tubabu ke d'o tirindi, Jaabe and the *tubab* asked if we had any land where we could grow rice. Our president, Musa Bacili, said yes. We showed them our ponds; they said there was too much water there. Then we showed them the land we're farming now. At that time there was nothing there but bush and small game. We cut down the trees.

Jaabe So

Amerikeni yogoni ri, some Americans came to Kuŋani. They said they wanted to help us farmers. I said: 'On what conditions?' They said no conditions; they just wanted to give us things. They gave us fourteen million CFA. I told Robert to keep the money in his account.

We went to visit a Corsican who had a garden near Kayes, in Mali. He told Robert: 'You're lucky to have a chance to work directly with *paysans*. But watch out for SAED.'

After that, I learned that during a trip to Dakar, Robert visited SAED's headquarters in Saint-Louis. I told him: 'Robert, don't bring those people into it; *ce sont des serpents*.' He said he just wanted some information about rice-growing.

We went to Dakar together. On the way there, Robert said to me: 'Let's go to SAED headquarters together; if SAED wants the money to be given in its name, we won't sign. We'll fight for our fourteen million.'

We arrived at SAED headquarters at about ten. We parked the car, and he said: 'Wait a minute; I'll just go and see if the Director is in his office, or at home.' He went off; he was gone until noon. When I went to look for him, I got as far as the door; the *planton* said: 'You can't go in, there's a meeting going on.'

When he came back, I said: 'Robert, that's not right. We came here together, and you went off to meet them on your own. From now on, I won't be able to believe anything you say.' He said: 'I'm sorry; the meeting had already started when I arrived, and they said it wouldn't take long. They said we should just order what we want.' I said: 'You should have sent for me.' He said: 'I didn't want you to lose your temper.' We went back to Dakar to order some pumps. He would suggest first one thing, then the other. I could see he was playing for time.

Manayeli Jallo

Koota yogo, one day when we were working together, Robert told us about SAED. At that time SAED hadn't come here, and we knew nothing about them. The first SAED person we saw here was Niang; Robert brought him here to show him our fields.

Jaabe So

Rober yille katta Dakaaru; Robert went back to Dakar. The evening he came back, he said: 'We'll be having a visitor soon.' I said: 'What visitor?' He said: 'The Minister of Rural Development, it seems.' I said: 'The Minister, here? What does he want?' He said: 'Apparently he's heard that you're well organized here; he's coming to see.' I said: 'Is that so? What are we supposed to tell the man? What's he after?' He said: 'Nothing, he's just coming to have a look.'

> AID's initial identification of Bakel irrigation as a potential project was made by a team of consultants from Washington who traveled through the Sahel in late 1973 looking for activities to finance under the special AID Drought Recovery and Rehabilitation Program appropriation.
>> (Development Management in Africa, USAID (United States Agency for International Development), December 1985)

> My aim is to instil in the population the techniques and attitudes of mind required for the grandiose task of developing the Senegal River Valley.... CIDR works in close collaboration with official bodies such as SAED.... In the near future, the area's main crop should be rice.
>> (Speech made by Robert Aprin in Bakel in December 1974, to welcome the Minister of Rural Development)

Well, the Minister of Rural Development came and had a look. The vegetable garden was beautiful then: tomatoes, onions, cabbage. When they came back to the village, I was standing next to the Minister; he didn't know me, he only knew Robert. He said: 'I must say, these people are hard-working. I only wish it were for our benefit.' The Director of SAED, Cheikh Cissokho, was standing nearby. He is a prominent Bakel politician; we call him Gawusu. He said: 'Minister, why not?' They were talking among themselves.

The Minister said: 'Well, in that case, we must try and find something for the technician.' The Director said: 'Let's have him leave here and move to Bakel. I have a house there.'

I heard them say that. But my idea was that Robert was a man; it wasn't they who brought him here, so they couldn't make him leave.

They left. Next day, Robert said 'Jaabe...', speaking as a child does to its father, ashamed because he knew he shouldn't be saying it: 'Jaabe, they've found me a nice house in Bakel, and there's a market there. So I'll be moving to Bakel.'

I said: 'Will you? On what terms? It wasn't the Minister of Rural Development who brought you here. We said that at the end of the season we would build a house for you by the river, with plenty of space, and a garden and all. You agreed. Now you tell me you're moving to town. What's it all about?' He said: 'No, you'll see, I'll still come here.' Well, I couldn't stop him; but I could see what they were up to. So he left us and moved into Bakel.

> Another ambiguity, calling into question the 'development volunteer' spirit often connected with NGO agents, is the fact that the expatriate technician lived under the roof of the chief initiator. That presence undoubtedly gave added stature to a person who was already far ahead of the rest of the population in his thinking about rural development. To enhance in this way the influence of a person already in control of the situation, while aiming for a form of development based on equal participation for all, was not very logical.... Finally, as it was unthinkable to spend years living 'at grassroots level', in conditions difficult for a European family to endure, the departure of the expatriates for the nearby urban centre, only a few kilometres away, was diversely interpreted; by some in an uncomplimentary fashion, to say the least.
>
>> (Robert Aprin's thesis, 'Development and Peasant Resistance: The Case of the Soninke of Bakel, Senegal', 1980)
>
> At the time of my visit to the project in January 1975, the CIDR project Head had applied to the USAID office in Dakar for funds to be used for the purchase of irrigation material.... War on Want stressed to CIDR the numerous dangers at a social and economic level inherent in the unplanned introduction of motorized irrigation.
>
>> (Report by Richard Elsner of War on Want, co-funders with Oxfam of Robert Aprin's presence in Kuŋani)
>
> In Bakel, as in Matam, our task is to educate the farmer, so that when the dams are built he will be in a position to recycle himself without difficulty. Before that time comes, we must accustom him to the rigour and discipline required for the rational cultivation of rice, to which he must turn once the dams are in operation.
>
>> (Le Soleil, a Dakar newspaper, quoting the Director of SAED, 5 February 1975)

Modibo Koyita Tanjigoora

Rober da xasu seegi ña yere; Robert spent eight months here, in Jaabe's house. He didn't need to pay anything, because he'd come to help us with our work; it was for our sake he was here. We're poor people, we couldn't pay him; but we did everything for him that we could. After he left, two days would pass, then three, then a week, then two weeks, without our seeing him. Luckily, we'd managed to learn a little ourselves. If someone comes to teach you something, it's best to keep your eyes open, so that if he should leave tomorrow, you can do it for yourself.

Jaabe So

N do Gawusu da me ñi; I had a meeting with Gawusu. He said we should work with SAED; they would organize us. I asked what that meant, and he said they would teach us how to grow rice, apply fertilizer, and so on. I said: 'You must respect the farmers' decisions. They aren't troops; no one is paying them. They're farming to live.' He said that would be no problem.

Richard Elsner from War on Want came to visit us. He said they would withdraw from the project and stop paying Robert. Robert said: 'If I don't go along with SAED, they'll throw me out. I'll pretend to go along with them, but I'll really be on the farmers' side.' Richard asked me what I thought. I said: 'Robert's been playing their game for some time now. He went after them of his own accord. A competent man, a man with dignity, wouldn't have done that.'

While we were meeting with Richard, Robert said he needed a Landrover. Richard said he needed them for another project. Robert asked me to ask Richard again myself. Finally, Richard let us have a Landrover. When Robert went over to SAED, he took it with him. They painted 'SAED' on the door instead of 'War on Want'; and we had no vehicle of any kind.

Before he moved to Bakel, Robert told me there was too much work for him to cope with alone. I said: 'Why don't you write to Uccelli?' He said: 'No, if I write to Uccelli he'll recognize my handwriting. Someone else should do it.'

At that time, I hadn't yet realized what they were up to. So I told my son to write to Uccelli on my behalf, telling him that there was so much to do now that Robert couldn't manage on his own; could we have two more people? Uccelli wrote back: 'Why don't you hire an African technician?' Robert said: 'No, it has to be a European; *il faut un Blanc.*' So my son wrote to Uccelli once more; Uccelli wrote to me, accepting. I showed Robert the letter and said: 'Now you can write back yourself.'

Robert had two friends who needed jobs; he wrote to Uccelli asking for them to be sent here. That was how Lucien Gorvan and Robert Laborde came here.

Adrian Adams

I first came to Kuŋani in February 1975. It was not my first time in Senegal; I had lived in Dakar from 1963 to 1966, and taken my first degree there, but had not thought ever to return. After completing my studies, I had taken a post at the University of Aberdeen, in

Scotland. Unexpectedly, I found a strong interest there in African studies.

From the early 1970s, African migrant workers had been in the news in France; the Union Générale des Travailleurs Sénégalais en France, founded by Sally Ndongo from Sinthiou-Garba in the Middle Valley, was active on their behalf. I began to want to learn more about the African dimension of labour migration from the Senegal River Valley to France; and had decided to concentrate on the then less well-known Bakel area. I obtained leave from the University of Aberdeen, and took up a six-month grant from the Social Science Research Council.

Shortly before leaving for Senegal, I had met the historian Abdoulaye Bathily, whose family is from Tiyaabu. He gave me letters of introduction, among them one to Papa Kane from Golomi, then Director of the Ecole Nationale d'Economie Appliquée in Dakar. At ENEA, I met the students who had just returned from a *stage* in Kuŋani. Papa Kane gave me further letters of introduction, among them one to Robert Aprin. I met Robert at the station at Kidira, where he was meeting his colleague Lucien Gorvan. When I told him that I did not wish to stay in Bakel, he suggested that I ask to stay with Jaabe So in Kuŋani. I did.

One day Robert came to Kuŋani to explain that farmers' groups would have to come under the authority of SAED, in order to receive the pumps donated by USAID. He explained that this was a mere formality, required because he, Robert, was not entitled to receive such gifts in person, not being a Senegalese citizen.

Modibo Koyita Tanjigoora

Rober t'o na ro Sayedin wure; Robert came to tell us we should be under SAED; although he had previously agreed that we shouldn't be under the government, nor go into debt, but work for ourselves and slowly save up to buy the equipment we needed. *A da konbin dabari*; he plotted with SAED for them to come and take control. We didn't want them to, because we knew how that kind of thing works out, in the groundnut-growing areas or elsewhere: you go into debt, and then you have to sell your harvest to pay off your debts. We don't want debts. We just want freedom. *O ku ku, o na liberte baane*.

Adrian Adams

A meeting was held at Jaabe So's house in Kuŋani towards the end of March 1975; convened by N'Diaye, a SAED representative from

Saint-Louis, to explain their plans for intensive rice-growing in the area and how farmers were to fit in. Also present were Abdoulaye Diop, the organizer of a farmers' association engaged in irrigated rice-growing in the Delta; Robert Aprin, seated next to N'Diaye and Diop at the table, facing the assembled farmers; and a French sociologist, and a Chilean geographer, both under contract to do research for War on Want, who sat near the table but slightly to one side. I was invited to sit with them, but declined. Before the meeting, the sociologist remarked to me: 'Of course, Jaabe So is a real *kulak*.'

Several spokesmen for the Kuŋani group, Modibo Koyita Tanjigoora, Manju Jeba Tanjigoora, Mahamme Kamara, Basiru Tanbadu, explained through an interpreter that they wished to remain free to organize their own work. Robert came to help them develop their country, and they started work on their own. Then he came to tell them they should work with SAED. If what SAED wanted was really in peasant farmers' interest, why didn't SAED approach them directly?

Jaabe So said: 'We know SAED, make no mistake about that. SAED is Senegalese, just as we are. Robert isn't Senegalese. You should deal with us direct. We're not parcels to be handed about. *On n'est pas des paquets*.'

N'Diaye replied: 'You don't understand. Robert is now the direct representative of SAED. He has been named *président de périmètre* in Bakel. He will do all that he can to help you, whether as CIDR technician, as agent of SAED, or as just plain Robert.' Diop added: 'You're lucky to have Robert. You'll be able to make a good start. If you work hard, rich people will come and help you.'

Jaabe So said: 'Yes, we were lucky to have Robert. But Robert came here for development's sake, *pour faire le développement*. Now we hear he's an agent of SAED.' Diop said: 'Robert wants development for you. If he's collaborating with SAED, it's because he thinks it's in your interest to do so.' Jaabe So said: 'If Robert has our interests at heart, he should explain things to us.'

Towards the end of the meeting, another of the Kuŋani farming group leaders spoke: Al-Haji Isa Faatuma Silla Tanjigoora, a noted Koranic scholar. He asked whether people would be forced to join SAED. He was told in reply that Senegal is a democratic country. He then said: 'Ever since the world began, we've lived by farming. Our country wasn't built yesterday; it's an old country. God will help us to live. If SAED can make the sky fall, we will join SAED.'

N'Diaye protested that he was one hundred per cent Muslim. Diop

Hope Traduced (1973–1975)

> The collective farming project received a visit from the Minister of Rural Development and Hydraulics. Since that visit, the farmers of the Bakel area have realized that they are no longer alone, that they're being taken seriously, and that new prospects lie open before them. This visit has encouraged them to set out bravely on the road to development.
>
> (Robert Aprin's annual report on the work of CIDR, 1975)

> We had difficulty making the people of Kuŋani understand that we were students, come to learn about the life of peasants in the area. As far as we could tell, they considered us like *fonctionnaires*, who make suggestions but never solve anything....
>
> In spite of all the handicaps listed above, Kuŋani has potential for development... CIDR's work in the field of agriculture seems likely in the long run to destroy this conservative, highly stratified society.
>
> (Final report by students of the Ecole Nationale d'Economie Appliquée (ENEA), Dakar, who spent the last six months of 1974 in Kuŋani)

went even further. His activities, he said, were based on the Koran. Peasants were impervious to *conscientisation* and *vulgarisation*; God said so. God had brought SAED into being. Not to collaborate with SAED, would be like a son rejecting his father's legacy.

Robert, silent until then, finally remarked: 'There's nothing official between SAED and me. I'm working with them because I think it's a good thing. I'll be leaving later on; if you're left on your own, you won't be able to manage. My only mistake was to act as a go-between. I should have asked the Director to explain things to you himself.' N'Diaye added that the people of Kuŋani were very ungrateful to Robert.

Jaabe So

N d'a mugu Sayedin botu, it was at a meeting called by SAED that I heard that Robert had become an agent of SAED. That shocked me so much... Even now, I can't stomach it.

I said to him: 'That's not right. I've been telling you for a long time now that our development is a peasants' development, *un développement paysan*. We wanted a technician for ourselves; why did you have to go to SAED?' He said: 'If I hadn't joined them, they would have made me leave the country.' I said: 'If you were a man like me... Even if they threw me out, I wouldn't go along with them. If

I'm paid to help peasant farmers, I'll stay with the peasant farmers. Now you belong to SAED; you're not with us any more.'

Well, there we were. He didn't help us any more. We no longer had a technician.

Manayeli Jallo

Sayedin ga da tu, when SAED realized that Jaabe's work was going well, they said they wanted to join in. Jaabe said that he was a poor man, and had brought this work for poor people. If we were to work with the government, we would soon be in trouble. They would bring paper business into it: sign this, come here, go there. He wanted no part of it.

Adrian Adams

In May 1975 a second SAED delegation, larger and higher-ranking than the first, held a meeting in Kuŋani to announce that the government had drawn up development plans for the area; peasant farmers would be organized and trained (*encadrés et formés*).

Members of the farming group repeated that they did not reject equipment or technical advice, but wished to retain freedom of organization; after all, they started this work on their own. A member of the official delegation accused Jaabe So of being against the government. Another stated that no development was possible without the administration; Jaabe So could not have gone to France, nor Robert and Lucien come to Bakel, had the administration not allowed it.

Robert, present, remained silent. A few days later, Lucien came to Kuŋani to tell Jaabe So that during a meeting of local government officials in Bakel, the CIDR technicians were accused of encouraging the people of Kuŋani to oppose SAED, by letting them think they had a choice. Lucien suggested the possibility of dialogue, but the *préfet* said that the peasants weren't ready for it: '*Les paysans ne sont pas mûrs pour le dialogue.*' Henceforth, Robert and Lucien were told, they were to consider themselves agents of SAED; CIDR was out. Whenever they travelled outside Bakel, they would be escorted by a government official.

Jaabe So

N d'in Sanba-Salun ñiiñen kin'i ya, I gave my land at Sanba-Salu to the *groupement*. When we had cleared it, we wanted more land, so I asked Dawuda Haawa Tanjigoora to let us have the land next to mine, that

Hope Traduced (1973–1975)

On May 14, two of the expatriate technicians, travelling by car from Dakar to Bakel, were stopped at Saint-Louis. They were informed that a 'heavyweight SAED team' had gone to the Soninké area. They were forbidden to travel to Bakel before the SAED team returned. The Director of SAED stated that he was 'extremely displeased' with the CIDR agents, because they 'did not support' SAED agents working in the Bakel area, and their behaviour 'encouraged the peasants to subversion'.

> (Robert Aprin's thesis, 'Development and Peasant Resistance: The Case of the Soninke of Bakel, Senegal', 1980)

The CIDR project... was ordered by SAED to integrate its activities within government priorities; the project Head was nominated a government representative and from May 1975 began to take his orders directly from SAED headquarters.

> (War on Want report, October 1975)

AID was asked by the GOS [Government of Senegal] to consider assistance to a number of locally organized village collective groups. The Compagnie Internationale de Développement Rural (CIDR) has placed 3 expatriate volunteers in Bakel, under the auspices of SAED, with the hope of organizing small-sized irrigated perimeters in 15 river-side villages. The expatriates have succeeded in instilling a cooperative spirit among the population of the villages... AID has recently granted $60,000 for the purchase of 3 water pumps to be used on 3 perimeters which are to be in place by July 1975.

This project was developed after consultation with the CIDR volunteers and SAED representatives. It will consist of the preparation of surface irrigation systems to provide water for 1,320 hectares net of land in 15 perimeters.... SAED, as the implementing agent, will have to redesign its structure and strategy as it extends into the upper basin, to be more sensitive to and make greater allowances for sociological factors which should be regarded as a needed catalyst in assuring the success of this undertaking. AID's experience in this area can provide a basis for a constructive and critical stimulus which should be regarded as a needed catalyst in ensuring the success of this and other donor activities in the basin, and to give maximum encouragement to initiatives springing from the population itself.... The CIDR undertaking, the pilot program for this project, will also make an important contribution. The CIDR agents have introduced the cooperative spirit as well as the fundamental farming techniques. The contacts they make with the farmers in the villages, the results of their pilot projects, will set the course for any future endeavours....

> The estimated cost of the AID financed portion of the project is $3,100,000.
>
> *(USAID Project Review, Project Number 685-0208, May 1975)*
>
> Encadrement *(project management)* is at present ensured by CIDR (3 expatriates); it is necessary that this assistance be maintained.
>
> *(SAED application to USAID for funding for the 'Bakel Small Perimeter Operation', July 1975)*

had been given to his father's elder brother Muusa Kaba. He refused at first. But I talked him into it, saying that as he too was a member of the farming group, he ought to give his land as I'd given mine.

In the beginning, it was his land and mine the group farmed. It was afterwards that the *debegume* said that the best thing would be to give all of Sanba-Salu to the *groupement*. The *Bundunko* household objected at first, as they were using their field at the time; but Al-Haji MaJaaxon Ba decided they should accept the town's decision. That was how the *groupement* was given the land at Sanba-Salu.

Modibo Koyita Tanjigoora

Yittu xooro, there were tall trees at Sanba-Salu. We cleared the land and burnt the tree-stumps. Afterwards, they brought a bulldozer. We built the canals and dikes ourselves, with picks and shovels.

Jaabe So

I da mesin riti, they brought a pump and installed it at Sanba-Salu. That was the first pump in the Bakel area, not counting mine. The day they brought it here and started it up, I said: 'You say this is a new pump, that's never worked? Listen to the sound of the motor.' They said I was crazy.

When work began, we found bullets in the earth from the time of Seexu Mammadu Lamiina. My father's first wife, Juumu Haaruna Moodi, once showed me some trees at Sanba-Salu, five *jebe* and a *sexene*, saying that at the time of Seexu Mammadu Lamiina's war, people climbed them to look out for enemy troops. When the bulldozer came to prepare the land for irrigation, they wanted to uproot those trees. I said no, and they left them standing; but after a few years, they died and fell. There are still some *jebe* in the *kolangal*, though, that were there in Seexu Mammadu Lamiina's day; they are stronger than the younger trees.

Adrian Adams

When I left Kuŋani in July 1975, Jaabe So asked me to write down the true story of what had been happening there.

6

Behind the Lines
(1975–1978)

Year Three: 1975/1976

Jaabe So

Kuŋani da maaron soxo, the Kuŋani farming group grew ten acres of irrigated rice and five acres of rain-fed sorghum at Sanba-Salu, and two and a half acres of rain-fed sorghum at Fanqanne. We organized six teams of men and six of women. Every week there were three days of collective work, all the teams together. On the other days, one team of men and one of women went to work in the fields.

Adrian Adams

There were 224 people in the group, 94 men and 130 women. Almost all the households of the town were represented, some by one person, some by two, some by three, or eight, or twelve. The heads of the men's work-teams were Basiru Tanbadu, the son of Al-Haji Abdullayi Tanbadu; Sanba Dukure, whose grandfather accompanied Seexu Jomo back from Gori; Tijani Natoxoma Tanjigoora, Manju Jeba Tanjigoora, Modibo Koyita Tanjigoora, and Isa Tamura. The heads of the women's work-teams were Saajo Tanjigoora; Xujeeji Kebe, married into a Tanbadu household; Xujeeji Dabo, married into Sanba Siise's household; and Naye Njaayi and Ganciri Jallo, Mahamme Kamara's wives. Fenda Sidibe, married into a Jarra household, headed the twelve women who sold vegetables in season. The storekeeper was Siliman Sanqare; he was also one of the pump operators, along with Al-Haji Mpali Tanjigoora, MaJaaxo Isa Tanjigoora, and Mammadu So. Dawuda Haawa Tanjigoora, the son of Muusa Kaba's younger brother, was responsible for discipline, roll-calls and fines. The treasurers were Al-Haji Isa Faatuma Silla Tanjigoora and MaJaaxo Isa Tanjigoora. The secretary was Siixu Jallo. The *chef de chantier*, present in the fields every day, was Mahamme Kamara. Jaabe So was president, and the women's president was Maamu Sakiliba, Jaabe So's first wife.

Al-Haji Mpali Tanjigoora

Gilli o ga d'a joppa, from the beginning Jaabe always said the same thing: whatever else we did, we should farm a collective field. We agreed on the rules we would follow. If you were late on collective work-days, you paid a fine; if you missed work altogether, without a reasonable excuse, you paid a larger fine.

Jaabe So

Mesin ga na golliña lenki, if the pump worked today, it wouldn't work tomorrow. If it didn't work, we would run and fetch a mechanic from Bakel. It was an old Italian pump, that had been used in the Delta. It was just sitting in a SAED workshop, until they repaired it and painted it, and said it had cost eight million CFA. That was the start of *développement administratif*: nothing but trouble.

The rains were not good that year. Weeding the rice was very hard work, just when the rain-fed fields, sorghum and groundnuts in the *jeeri* and maize at Fanqanne, needed most attention. We harvested fourteen tons of rice. If we'd been able to irrigate properly, we would have farmed a larger area and had a better yield. The rain-fed sorghum did not do well; we harvested only a ton. The rice was sold locally at 50 francs a kilo, and the sorghum at 5,000 francs a hundred-kilo sack; the money was put into the group's *caisse*.

The other farming groups worked in much the same way. The Bakel group grew eight acres of irrigated rice, and the Baalu group twelve acres.

Jaabe So

It was during the 1975 rainy season that a group of Bakel *fonctionnaires*, including my nephew Bokar Sisoxo, sent a paper to people from the Bakel area working in France, saying that the government had brought a technician to help the farmers, and asking them to send money. They signed our names to that paper without telling us, at the very time when we were struggling against SAED. We saw that if things continued like that, the *paysans* would never become strong; SAED and the money it received from outside would divide us among ourselves. So we decided that all of us who wanted to develop our country should unite to form a Federation.

> Thanks to the Senegalese Government's determination to save the département of Bakel from its state of underdevelopment, CIDR, whose aims are essentially philanthropic, has been active in our area for the past 15 months, helping Soninke and Halpulaar farmers to form co-operative groups... SAED has already provided three pumps; a generous gift from the Senegalese Government, whose constant solicitude towards our peasant farmers no longer needs to be proven....
> Unfortunately, a large part of the millions of CFA francs you send home is kept hidden in suitcases, with all the risks that implies... Could these enormous sums of money not be invested?...
> We inform you that following upon the creation of numerous farmers' groups, our intellectual cadres, sons of the département, have formed an advisory committee... Those of you who are interested in the projects we have mentioned, can contact this Committee.
>
>> ('Letter to all migrants from the département of Bakel living in France', Bakel, July 1975)

Mammadu Lasana Ba

O do Sayedin ga golliñaana, when we began working with SAED, we here in Jawara and others found that the work didn't go the way we wanted. Afterwards, the farmers themselves said: 'Let us organize and make a Federation, all the *groupements* together.'

Jaabe So

I first heard the name 'Federation' when I was on a voyage to Madagascar. There were both Whites and Blacks growing sugar. In Réunion, we went to visit a sugar plantation. We went to a big store; there was a girl there. She stood at the door of her office, with a pen in her hand, and I chatted with her. She asked me where I was from, and I said: 'Marseilles.' She said, 'Are you from a ship?', and I said: 'Yes; the *Ville de Massanga*, Compagnie Worms.' She asked, 'Why have you come all the way out here?' I said, 'I've come to see the sugar-cane fields.'

She said: 'Further along the road is a big plantation that belongs to Whites. This plantation here is our own; it belongs to people from three towns. The work only started to go well when we formed a Federation. Once we'd formed a Federation and registered our statutes, we had the right to make brown sugar and sell it ourselves; when ships come to take the sugar away, they take ours as well. If three ships come, there's a White here who takes one all for himself. The people from the government plantations take the two other ships. The one or two ships that come after that, are ours. That's how we won respect for our work. We have a lot of money now. It was difficult for us to organize, but there's a Creole here, with a Black mother and a French *douanier* for a father, who has a good job with the government; he helped us with our statutes, and he's still helping us now.'

That was when I first heard of a Federation, and it stayed in my mind. After that, I found there were some fishermen in the Vieux Port in Marseilles, who only dealt with crayfish; they'd formed a Federation with some fishermen in Spain to sell their catch together. Their price was lower than the market price, and hospitals and the Army bought from them. So I understood that a Federation was a good thing.

When we held our first meeting in Bakel, that rainy season, someone said: 'What's a Federation?' I said: 'If the farming groups in Manayeli, Tiyaabu, Jawara, Mudeeri, and so on all join together to make a large group, that large group is called a Federation.' It was at that meeting that we decided to form a Federation, in order to keep our independence as much as possible. Not everyone understood it; even today . . . But I knew that it was important.

During that dry season, the Kuŋani farming group grew irrigated maize and vegetables at Sanba-Salu; we abandoned the well we'd dug at Fanqanne the year before. Robert came to see us, and said we should just sow the maize straight away, without making furrows. I said: 'I know they sometimes do that in France, but that's for overhead irrigation; what we have here is furrow irrigation.' He insisted; so I said we should divide the maize-field in two, and sow half with furrows, half without. After three weeks, you could tell the difference; Robert agreed I was right.

Richard decided to stop paying Robert's salary. Robert became *chef de périmètre* at Bakel, and stopped coming to see us in Kuŋani. Lucien supervised the farming groups downstream from Bakel, and Robert Laborde was responsible for the groups upstream from Bakel, including Kuŋani.

Richard came to start a new project on the other side of the river. I heard of it when relatives and friends from there came to ask me who this *tubab* was, what he wanted. I told them: 'He used to be our financier. I don't know what he has in mind for you; why not ask him yourselves?'

Adrian Adams

War on Want said in London that they still wanted to help the Federation; perhaps through their Chilean geographer, who would be going out to work on the new project.

From the USAID we learnt that a major development plan that was passed by Congress in the spring, devoted 3.1 million dollars for 'small irrigated perimeters' to be situated in the villages where CIDR had done its animation work... This first 3 million dollar contribution is seen as the first stage in a major 8,000 hectare large perimeter production scheme which would take effect in three to five years. In no sense could these plans be presented as unimportant; they are sure to change the whole nature of the economy of the area, and in ways which War on Want believes will lead to a very impoverished form of development. However USAID stressed the importance they place on the CIDR team which they consider to be essential to their long term development plans for the region... Long discussions were held with the CIDR project team, with particular reference to their own objectives and those of SAED. War on Want's conclusions were that, contrary to the team's beliefs, the team had limited ability to influence the plans of SAED and USAID... The social and economic effects of irrigation were sure to invalidate the previous objectives of the project, namely the raising of the productivity of traditional agriculture.

War on Want proposed that the team re-establish a peasant-based project orientation, given that now the team had departed from the original project's objectives and methods under the pressure of SAED. We considered this to be the only viable alternative but of course we could not enforce it. We left the choice to CIDR, and they have decided now to implement the government's plans... In the months following SAED's takeover of the project, considerable tension developed between the CIDR team and the local population who came to see that the team's technical advice was being prompted by SAED. It remains to be seen exactly how the villagers will react to the increasingly severe control of SAED over their production, but it is clear that the position of the CIDR team between SAED and the population is increasingly untenable. It follows that the team will have to commit itself openly to SAED's imperatives which we believe will lead to the alienation of the population from the team.

(War on Want report, October 1975)

Jaabe So

Kineyen waxati, during the dry season there was a meeting of farming groups at the *mairie* in Bakel. There were nineteen groups, thirteen from by the river and six from the *jeeri*. We decided to make a Federation and draw up its statutes.

They said I should be president, because I'd brought the work here.

I said: 'It's hard for me to refuse. But I'd rather just farm my land at Seegankaani. If I become president of the Federation, I won't be free to do my own work.' They insisted. Saajo Fenda Tarawore, the president of Bakel-Gassanbilaqe, said: 'We're behind you, you're in front of us. Wherever we die, we and you will die together.' So I accepted.

I said that as the *Fan-lenme* and Bundu were a long way away, I would like to have Seyidu Ñaŋaane as my vice-president. No one was pleased with that. But I said that he was a younger man, he'd be at hand for me to summon when I needed him. So they agreed. Then we took Ali Tarawore, a retired soldier from Bakel, as secretary. We chose a second vice-president. We chose two treasurers, and a *commissaire aux comptes*. We carried on with our work. For my part, I accepted the burden once and for all. As for Seyidu, the whole world has seen how he turned out.

Bokar Sisoxo wrote our statutes. It was he who proposed the name *Fédération des Paysans Organisés en Zone Soninké de Bakel*.

Mammadu Lamiina Bacili

Ken siine, by 1976 Jaabe had formed the Federation. Our president Musa Bacili attended the meetings and told us what happened. At that time, Jaabe was president of the Federation, and Seyidu Ñaŋaane was his vice-president.

Modibo Koyita Tanjigoora

Sayedinko ri, the SAED people came and said we should buy 200 kilos of fertilizer per acre. But Jaabe So said we ought to find out how much fertilizer is really necessary. It would be best to know exactly what we need, fertilizer, fuel, seed; otherwise, they'd just tell us: 'Take this, take that.' Then we'd have debts hung round our necks, we wouldn't be able to pay, and SAED would be able to tell us: 'Do this, do that.' We decided that Siixu Jallo and I should go to the Lower Valley to find things out. So we went, and saw how they worked. They told us: 'Your soil is good, it's not been used much yet. It doesn't need much fertilizer; say forty kilos per acre. We came back and explained what we had seen. We worked out how much fertilizer we needed for the two seasons. In that way, little by little, our eyes were opened.

Jaabe So

A su yi tu, everyone knows that when you offer credit in Africa, people take it as a free gift. SAED won people over by offering to deliver fuel and fertilizer right to people's houses, all on credit. The work went wrong from then on, and many people ended up leaving.

I was asked at the time to say what I thought of SAED. I said that we weren't against SAED. But we'd seen that SAED didn't explain anything to the farmers; it just drew them into debt. At the start of the season, SAED said: 'Here you are, you need so much fertilizer, so much this, so much that.' At the end of the season, it said: 'Now you owe me 400,000, 500,000.' Even if the farmer sold his whole crop, he couldn't pay the 400,000; he'd still be in debt. Next year, he'd say: 'Right, I'll pay those debts.' After the same business with fertilizer and all that, he'd owe another 400,000. That man couldn't ever be free. In the end, he'd just want to be rid of debt, to get out of development. That's what we were against.

SAED didn't give anything away. SAED took advantage of the peasant farmer, and put him in debt: on fifty acres, seventy-five acres, there would be 500,000 francs owing. What about the farmer's work, the ten or twenty people in his household? Could that man make ends meet? He couldn't. He couldn't. That's when people gave up. When SAED came into River development, people left. People who saw things clearly left, and development was bankrupt.

SAED worked for SAED, not for the farmer. We knew that, because we saw what went on in Matam. We wanted to work as independent peasant farmers, work in a small way and earn something for ourselves and our families.

So we knew about SAED. SAED was the government, and we were peasant farmers. SAED was here; but peasant farmers also needed to live. We had a right to live, like everyone else in the world. That's why we wanted to develop. If that point were lost, then development would be worth nothing.

Year Four: 1976/1977

Jaabe So

We cleared twenty more acres of land at Sanba-Salu; that made thirty acres in all. An American bulldozer cleared the trees. We built the canals ourselves, by hand.

We'd trained two pair of oxen the year before, for the whole group to use. We entrusted them to Mahamme Kamara's son Siixu, along with an ox-cart, saying that he could keep whatever he earned with the cart, and whenever he ploughed the collective field for two days, he could plough two days for himself. But it didn't work, because on days when he was meant to work in the collective field, he would only work for an hour or two. So we took back our oxen and cart, and took turns ploughing with them. But oxen need to be fed and watered by one person. So we gave up the group's oxen. Basiru Tanbadu trained a pair of oxen for his own use, and I trained a pair; we lent them to the group when needed.

When Robert Laborde came, I said: 'Right, the land is ready, what shall we do now?' He said: 'Just sow the rice.' We sowed all thirty acres. When we'd finished, I said: 'What do you suggest we do now?' He said: 'Guard the fields, so birds don't do any damage.' We set nine people to keep watch every day, not including the *chef de chantier*.

The first rain came; it wasn't enough to make the rice grow. All that time the pumps were right there, and we didn't even think of using them; we didn't know. The technician didn't say anything. The weeds began to grow.

Eight days passed between the first and second rains. The rice sprouted with the second rain. By then, the weeds had already come up; you had to search among them to find the rice plants. The technician came and said: 'You'll have to get rid of the weeds.' I said: 'What do you mean, get rid of the weeds? How can anyone get rid of those weeds?' He went away.

I called a meeting the next day. I said: 'All we can do now is give up that field and sow another one.' We sowed another ten acres of rice. Well, some people from SAED came to visit. They said: 'Why was that rice taken over by weeds?' Robert Laborde said: 'It's the farmers' fault.' I said: 'Look, that's not true. We sowed thirty acres; you were there. When we'd done, you said we should guard the field, to keep the birds away. For a long time, it didn't rain. The pump was right here, and you didn't think to tell us to use it. Why not? Why don't you answer me, in front of these people here? We don't know about growing rice. Then it rained; by the time the rice sprouted, the weeds were eight days ahead. Even a fool could tell it was hopeless. I told you off; I said you were no good as a technician. You went off to Bakel. When you came back, you found that we'd tilled and sown part of the field again; we'd given up on the rest. Why don't you tell the truth?'

The SAED representative asked: 'How many days passed between the time you sowed the fields, and the first rain?' I said: 'Nearly two weeks.' He said: 'It's not the farmers' fault.' They found we were in the right; but the technician wanted to blame us.

They took Robert Laborde to Saint-Louis, then to Matam. They told him he should get us to divide up the collective field into individual family plots; but we refused. We said we could make each team within the group responsible for a specific part of the collective field, but we wouldn't give up working together.

I was spending all my time on farming group business; I didn't have time to farm Seegankaani any more. After that, for four years I didn't farm at all. Then, when I started farming again, it was in the group's fields at Sanba-Salu, not at Seegankaani.

The Kuŋani farming group harvested thirteen tons of rice. The rains were not good that year. We decided to give twenty kilos of rice to each group member; each member was also entitled to buy up to two hundred kilos.

Eight other groups started growing rice that year. SAED told people to grow rice on sandy soil; you would irrigate one day, and two days later you would have to irrigate again. People used a lot of fuel and fertilizer for nothing. In the end, I went round to all the groups, telling people that next year they should refuse to grow rice on sandy soil.

During the dry season, Kuŋani grew eight acres of maize, as well as vegetables. Out of our earnings we paid what we owed SAED for both seasons.

Siixu Jallo

One day out in the field, Robert Laborde said to me and Jaabe So and a few others: 'Experience has shown me that collective farming doesn't work. The best thing is to divide up the land into family plots; each family will work its own plot according to the schedule drawn up by the technician.'

Robert moved away from Jaabe So and approached me. He said: 'You know I'm right, but you always agree with So. You're the most highly educated person in the village; try and convince the others.' Then he said to Hamme Kamara: 'You're just afraid of So. He wants to be boss, that's all. In any case, we've got to get things straight before January 1977, because if next year is like this year, the Americans will fuck off and go fund something else; they only fund things that work.' The discussion ended with those striking remarks of his. Thoroughly vexed, he went back to Bakel.

Jaabe So

Ken falle, afterwards Robert Laborde suggested we go to a meeting in Matam. They killed a sheep for us; they showed us individual family plots everywhere, to show us that *l'individuel, c'est bien*. The *tubab* in charge at Matam told me, 'So, we've brought you here to show you that people do well out of working individually, each for himself. Just look around; talk to the farmers.' I said: 'All right, we'll talk to the farmers.'

We looked at everything; afterwards there was a meeting. The Matam technician talked and talked. When he had finished, one of our people spoke; then I did. I asked the Matam people if they had a common fund, a *caisse*. They said no. I said: 'Explain how you can have a development without a common fund. *Un développement sans caisse, ça m'étonne.* How did you do it?' They said: 'SAED came; they brought machines to make the land ready, then they divided it and gave each person his own plot.'

I said: '*Ce n'est pas un développement paysan. C'est un développement administratif.* It's interesting for us to see how you work. But it's clear that we're not organized in the same way. We're organized peasant farmers, we're working for a peasants' development, in the interest of peasants. What you have is administrative development. You should come and visit us, to see how we work together. We have a common fund, to give us strength. We ourselves decide what we want to do. We ourselves decide how many acres we want to sow. Your work is nicely done, your irrigation canals are fine; but we're not interested in your style of organization.'

Straight away the Matam *tubab* was angry, he took Robert aside to say they should stop me talking, otherwise I would harm their work. Then he said: 'Fine, that's enough; why not leave it at that and go have something to eat.' He and Robert weren't pleased; they went off in the car, leaving us out in the fields with the Matam farmers. We asked them how much they made out of their work; we saw exactly how things were run there. It was no good. We returned to Matam; straight away, they sent us off to Bakel. After that there was no more talk of trips to Matam.

It was then that SAED tried to put a Kuŋani *moodi* in my place: Al-Haji Modibu Tanjigoora, Fode Mammadu's grandson, the head of the Koranic school at Xoje. He had some land at Sanba-Salu, and *taalibo* to work for him. They stopped coming to see us, or telling us anything;

The area is populated almost entirely by subsistence farmers who have survived since time immemorial on a marginal existence basis. An extensive sociological study was made to determine the motivations and effects of the project on the farmers. The sociologist found that the project was indeed changing the social structure of the area but that the changes were those which are needed to move the area into the modern society. He found great support for the project among the villagers and predicted wide acceptance of the project by the target group. He further found that the methodology used by SAED and technical assistance in the project area was the approach most likely to succeed. The Sarakolle are essentially very independent people and pride themselves on their ability to do things on their own. The basically 'hands off' policy of SAED is very appealing to them. . . .

This project is unique in two ways. First, it is already underway and the development potential has been identified and demonstrated in both the human and technical realms. Second, and even more important, the project is a direct result of actions initiated and requested by a resident farmer of the area. At the risk of sounding like a fairy tale we would like to present the story of the Kounghani development and how this moved into an area development program following a truly grassroots developmental process.

The project was begun at the initiative of one farmer from Kounghani village. During the time he was working in France and travelling in Europe, he noted the improved agricultural methods applied there and upon his return to Kounghani he bought a pump and a rotatiller to use on his land. Because of insurmountable technical and logistical problems (no mechanics, lack of fuel, no spare parts available, etc.) his equipment was not useable. Realizing the need for expert assistance he wrote to the Director of CIDR, whom he knew because he used to work in the same building in France that CIDR has its offices, requesting assistance.

CIDR sent its first man to Koughani in 1974. He and his wife lived in Koughani for several months learning the way of life and organising the first farmer group. He then moved to Bakel to begin activities in other villages. Two other CIDR volunteers arrived in 1975 to assist with the extension effort and SAED began contributing some assistance in the form of pumps and technical advice. The CIDR group moved into irrigation work in addition to traditional culture in 1975. As a result of this move to a more advanced technology, War on Want withdrew its support. In the meantime AID had begun to support the project with R&R funds, providing additional pumps and tools.

The SERDA group (formerly CIDR) has overcome the ever-present 'startup' problems inherent with the introduction of a new technology and has established a working relationship with the farmers of the area. This

Behind the Lines (1975–1978) 143

process, which is time consuming and not very spectacular took over three years to accomplish. AID's entry into the project now, when project needs are higher and the program is accelerating after having proven itself, is very sensible. . . .

(Project Agreement between USAID and Government of Senegal, 1977)

---

whenever they came here, *tubab* technicians, African technicians, they parked their cars by Al-Haji Modibo's door.

SAED gave Al-Haji Modibo three pumps in all. They gave him one without the float it needed; they gave him another with a small float; then they gave him another, a Lister HR2. All that so he would rise and the Federation fall. But it didn't work.

Adrian Adams

According to the 1977 Project Agreement between USAID and the Government of Senegal: 'AID through the R&R program assisted with financing the 1974 program, and in 1977 provided "pre-project" funding of $124,000 to keep the SERDA team in the field, thus avoiding a complete loss of momentum. The total AID contribution to date is $294,000 and financed pumps, tools, studies and costs of the CIDR/SERDA staff in 1976 and 1977.' (An expatriate member of staff at ENEA told me at the time that USAID money passed through SERDA, a Senegal-based consultancy set up by a high-placed government official, which took 5 per cent, and SERDI, based in Belgium, which employed Robert Aprin and Lucien Gorvan.) The original SAED request in 1975 was for $1,345,000 for a 500-acre pilot project. The Project Agreement was for $6,671,000 (1.667 million CFA at the exchange rate then prevailing) and 4,500 acres.

A study funded by USAID reported high rates of malaria and schistosomiasis infection in Kuŋani.

According to the authors, SAED claimed that the study's concerns were not valid, rejecting the suggested distribution of free chloroquine and asserting that only twenty-eight cases of schistosomiasis had been officially reported in the area.

Jaabe So

An economist from ORSTOM [Office de la Recherche Scientifique et Technique Outre-Mer] asked to come and live in Kuŋani, to study how

> *In Kounghani village, April 1977, a 15.8 positive test for schistosomes across all ages, and 50% infection rate for the 15–19 year age-group for males, and 66.7% for females of the same age-group. Kounghani village has a 39.4% rate of malaria infection, and there is evidence to suggest that introduction of the irrigated perimeter at Kounghani may have helped keep the malaria rate up, compared with Golmi, a nearby village which did not yet have a perimeter. . . .*
> *Among the benefits of the Bakel irrigated perimeters:*
> *(1) Improvement in local diet and attendant reduction in mortality and morbidity rates, possibly particularly among infants.*
> *(2) Improvement in the local standard of living, critical to education, sanitation and other improvements in life.*
>
> *(American Health Association Final Report, Environmental Assessment and Health Component, Design of Bakel Irrigated Perimeter Project, 1977)*

people spend the money sent by family members working abroad. Researchers come here asking things they wouldn't dare ask people in France. They think we're still asleep. When they come here asking what we do with our money, some people here say: 'This lot don't know anything. Soon they'll be asking us what we do with our wives.' Don't they know people need to live? No *chercheur* can understand this country to the depths. This is a living country. There are living people here. This river has been here since the world began.

Adrian Adams

Richard Elsner and War on Want's Chilean geographer visited Kuɲani briefly in early 1977. According to Richard, USAID made it a condition of their aid to SAED that collective fields be divided up into family plots, and only rice be grown. He asked whether the Federation still existed. About the new War on Want project across the river, he remarked: 'If we don't do it, we'll leave a void, and other organizations, American ones especially, will move in.' He also remarked in the course of conversation that CIDR (founded by the son of the founder of the French Fascist organization Croix-de-Feu, de la Roque) had lepers living at their training centre for agricultural technicians; a trainee's attitude to the lepers allowed them to judge how well he would get on with Third World peasants.

Jaabe So

Sayeti da botun dabari; SAED held a meeting in Bakel with local *fonctionnaires* and representatives of farmers' groups. Robert and Lucien were there. The *préfet* of Bakel said the population was creating problems for the Bakel project, by disobeying the instructions given at previous meetings with SAED. He asked Robert to tell people once more how SAED was going to work, and to explain the contract SAED wanted the *groupements* to sign. When I said that we in Kuŋani would rather introduce irrigation gradually, and choose for ourselves what crops we wanted to grow, the *préfet* said that was out of the question; the pumps were meant only for growing rice.

Robert left the country soon afterwards. Later on I had a postcard from him, from Saint-Rémy-de-Provence; he said he had bought an orchard there. For my part, I have always regretted bringing him here.

Year Five: 1977/1978

Adrian Adams

My book *Le long voyage des gens du Fleuve* was published in early 1977. By then, I had returned to Kuŋani. I had a grant from the Social Science Research Council of Great Britain, to study land tenure and land use in the Senegal River Valley.

In 1977 a hundred-odd men, belonging to seventy-three Kuŋani households, farmed 225 acres of riverside land under rainfall, at Fanqanne and Sanba-Salu, and twelve acres of flood-recession *falo*. Thirty-five men belonging to twenty-seven Kuŋani households (including fifteen men and fifteen households of those also farming riverside land) farmed 135 acres of *jeeri* land that year, at Papata, Sanba-Gawulo, Lugere, Gece, and Guñan Yaamadu.

In all then, 120 men of 85 households farmed about 375 acres of rain-fed or flood-recession land, which in a good year could have yielded at most 150 tons. The people of Kuŋani need at least 300 tons of grain a year. The rains were not good; the *jeeri* yielded only ten tons or so of grain. Over two hundred men from Kuŋani were working in France that year; most of them in the Paris area, living in hostels in the thirteenth and fourteenth *arrondissements*, and in Saint-Denis.

Almost three hundred Kuŋani married women farmed that year; that is to say all the married women of Kuŋani, except those who were too old, or ill, or away from home. Nineteen women grew maize on small plots lent to them at Fanqanne. All the others farmed *jeeri* land, nearly two hundred acres in all. They all grew groundnuts; a few also grew indigo and swamp rice, and one or two grew sorghum. If the rains had been good, each woman could have harvested three sacks of groundnuts, or five, or seven. The rains were not good; each woman harvested only a few *kande* of immature groundnut pods.

Jaabe So and his son Mammadu grew seven acres of maize at Seegankaani that year, two and a half acres under irrigation. They did not farm the *jeeri*. They sowed maize and cowpeas on their *falo* at Sanba-Salu, the *falo* the Bacili gave Mammadu Denba when he married Mariyaama Siise.

If the rains had been good, they could have harvested five tons of grain. Their household needs at least two and a half tons of grain a year. The rains were not good; but they harvested three tons of maize from their irrigated field. The *falo* yielded only cowpeas.

One of the three married women of the household farmed at Seegankaani that year; she harvested only one or two *kande* of immature groundnut pods.

Jaabe So, his son Mammadu, and the household's three married women belonged to the Kuŋani farming group.

That year, the *debegume* Amadi Geyi was still alive. His son farmed a small field at Fanqanne, two-thirds of an acre. They did not farm the *jeeri*. If the rains had been good, they could have harvested one or two hundred kilos of grain; and each household head would have given the *debegume* a *kande* of grain, amounting to about a ton and a half of grain in all. The rains were not good; they received nothing.

That year, one of his sons was living in Dakar; another was in Kaolack, and another in France.

Two of the household's five married women grew groundnuts that year, one in Bema and one in Jaabalu. They harvested only a few *kande* each of immature pods.

The son in France was a member of the Kuŋani farming group.

That year Mammadu Abdu Tanjigoora, Isa Budu, was still alive. Three men and six youths of his household farmed that year. They grew *gajaba* on the one-acre field at Fanqanne, high on the alluvial ridge, that has been theirs for five generations, their share of the land that was given to

Al-Xaali Madi Bidan when he arrived in Kuŋani. (The rest of that land was farmed that year by the household descended from Buna Fasunte.) Their second Fanqanne field of one and a half acres is on the slope that leads down to the seasonal pond called Habalu; it was given to their household when that land was cleared in the 1930s. They sowed *gajaba* on the upper slope, and maize on the lower. They also farmed two and a half acres of *jeeri* at Garsingide, part of the land that was given to Buna Fasunte. After the rains, they sowed maize and cowpeas on their *falo*, on the slope of the creek-bed that separates Sanba-Salu and Seegankaani, where Seexu Mammadu Lamiina's people fought the soldiers from the fort in Bakel. This *falo* was given to them by another Tanjigoora household, not very long ago; the other household, descended from Buna Fasunte, farm the *falo* he was given at Sanba-Salu.

If the rains had been good, they could have harvested three tons of grain in all. Their household is a large one, and needs at least nine tons of grain a year. The rains were not good; they harvested nothing from Fanqanne, and no more than a few hundred kilos of grain from their *jeeri* field. The *falo* yielded only cowpeas.

Eleven of the household's nineteen men are away from home: two in Dakar, three in Ivory Coast, six in France.

Fifteen of the household's twenty married women farmed Suxangide that year. They harvested only a few *kande* each of immature groundnut pods.

Six of these women belonged to the Kuŋani farming group. Three of the men had joined at the outset, but left after two years because they felt it was too much work.

Isa Budu

Ma lenki ga rini, as of today we are old and just sit here. The young go to Paris. The rain does not come now; if you farm, you harvest nothing. Our children go to work for the *tubab* in Paris, and send us money. We buy sorghum and rice, we buy oil, we buy meat and fish. We buy everything; that is how we live now. Our children feed us by sending us money. There is nothing here. The only work we have here is farming; if it doesn't rain, we have nothing.

We buy sorghum and put it in the storeroom; every day we measure out six or seven *muudu* of grain. Praise be to God, the whole household shares the same meal, young and old, men and women: a hundred people, not counting strangers. Just one person is responsible for feeding all those people; the children send him money, he buys grain and measures out each day's ration. *Xaalisi ga ma ri, o kalla de!* If the money did not come, we would die.

Adrian Adams

Fode Mammadu's son Al-Haji Haamidu Tanjigoora, an elderly recluse, was still alive and head of the Xoje household that year. Three of his younger brothers, Al-Haji Sanba Jomo, Mammadu Xunba, and Siixu Fanta, grew maize at Fanqanne that year with fourteen *xaralenmu*, on five acres of their own, two acres lent by the *debegume* Amadi Geyi, and one and a half acres lent by Fode Daraame. Xoje also farmed two *falo* that year, at Fanqanne and Sanba-Salu. Al-Haji Haamidu's son Al-Haji Modibo was head of the Xoje Koranic school; his *taalibo* sowed fifteen acres of sorghum in the *jeeri* that year, at Papata.

If the rains had been good, Xoje could have harvested four tons of grain; and Al-Haji Modibo, six tons. Xoje needs at least thirteen tons of grain a year. Al-Haji Modibo runs a separate household, and his *taalibo* board with townspeople; he probably needs about five tons of grain a year. The rains were not good; there was no harvest from Fanqanne, and Al-Haji Modibo had less than a ton from Papata.

Of the twenty-five men of Xoje, Al-Haji Haamidu's brothers and their sons, seventeen were away from home that year: one in Dakar, one in Tambacounda, five in Congo, six in Ivory Coast.

Twenty of the household's forty-three married women farmed at Dogo and Jaabalu that year. They harvested only a few *kande* each of immature groundnut pods.

No member of the household, man or woman, belonged to the Kuŋani farming group. Al-Haji Modibo's *taalibo* grew two and a half acres of irrigated rice at Sanba-Salu, just upstream from the farming group's field.

Sanba Haawa's son Mpamara Sanba Tanjigoora was still alive that year. One man of the household, his younger brother, with his son and a Malian seasonal worker, farmed two and a half acres at Fanqanne, lent to them by Amadi Geyi. They also farmed a *falo*, but did not farm a *jeeri* field. If the rains had been good, they could have harvested one ton of grain. The rains were not good; they harvested nothing. The household needs at least four tons of grain a year.

Mpamara Sanba spent twenty years in the French merchant marine, but had not been abroad since coming home after the war. A son of his was away that year, working as a seaman based in Marseilles; another of his sons, and a son of his younger brother, were in Paris.

Four of the household's eight married women farmed at Suxangide that year. They harvested only a few *kande* each of immature groundnut pods.

Dawuda Haawa Tanjigoora, the son of Muusa Kaba's younger brother, was still alive that year. He farmed alone at Fanqanne, with a Malian hired hand: three-quarters of an acre of his own land, and one acre lent by Jaabe So. If the rains had been good, he could have harvested half a ton of grain. The rains were not good; he harvested nothing. His household needs two and a half tons of grain a year.

Dawuda Haawa once spent three years as a trader in Gambia, and worked as a labourer in Dakar. His elder brother's son worked in France for some years; he retired early, with a small pension, after an accident at work.

Dawuda Haawa's wife Muna Tanbadu grew maize at Fanqanne that year, and groundnuts at Bema. She harvested nothing at Fanqanne, and at Bema only a few *kande* of immature groundnut pods.

Dawuda Haawa and his wife both belonged to the Kuŋani farming group; he was responsible for checking attendance and levying fines.

Dawuda Haawa Tanjigoora

O yi golliñaana pur xunbane; we are working for tomorrow. We are wearing ourselves out so that tomorrow other people, young people, can say: 'This is the work they started, let us join it.' That's what we are working for. If we had wanted to gain things for ourselves, to fill the house with grain or money—ah! that we have not achieved. That's not what we are working for.

The *moodini* who say that other people should work for them, and they in turn can say their prayers for them or help them enter Paradise, are liars. If you're studying and can't afford to pay your teacher, then yes, you can farm for him. But that ends when you finish your studies. One person cannot say another's prayers. What is good is for people to farm together, and save what they earn in order to move forward together.

Adrian Adams

Sanba Siise was still alive that year. He and his brother Denba, and three of their sons, grew maize on a small quarter-acre plot at Sanba-

Salu lent to them by Gayi Jabira, and farmed seven and a half acres of sorghum in the far *jeeri*, at Gece. If the rains had been good, they could have harvested three tons of grain. The rains were not good. They harvested nothing at Sanba-Salu, but two tons of grain at Gece. Their household needs four tons of grain a year.

Both Sanba and Denba Siise once worked in Dakar, Ivory Coast, and France. Five household members, three men and two youths, were working in France that year.

Six of the household's ten married women grew groundnuts at Dogo that year. They harvested only a few *kande* each of immature groundnut pods.

Both Sanba and Denba Siise belonged to the Kuŋani farming group, as did nine of the household's married women, including Denba's wife Xujeeji Dabo, 'Tuntunba', who first asked for women to be allowed to join.

Sanba Siise

O kisimani ñi soxono Papata do Guñan Yaamadu; our grandfathers farmed at Papata and Guñan Yaamadu. Now we have started to farm at Gece, because the rains are not good. At Gece, if it rains heavily the sorghum is drowned; you lose your whole crop. We spend the week out there. Now that there's the collective field at Sanba-Salu, we come back for the days of collective work, then go back out to Gece. The women also spend the night in the *jeeri*; they grow groundnuts. Each one returns to town when it's her turn to pound the grain.

Things are better now than before. Before, if you had no grain from the *jeeri*, you had nothing. Now, with what we harvest and what the children send, we have enough to live on. And with the collective field, we have things we didn't have before. *A toq'o kun ya maxa*. It's up to us.

Adrian Adams

Siliman Sanqare and his son farmed five rain-fed acres at Seegankaani that year, lent to them by Famaxa Sisoxo of Bakel, and a flood-recession *falo*. They did not farm the *jeeri*. If the rains had been good, they could have harvested two tons of grain. The rains were not good; they harvested nothing. Their household needs two and a half tons of grain a year.

Siliman Sanqare worked for seven years in Dakar as a labourer, just after the war; he later worked in France for ten years. That year, one of his elder brothers was living on a pension in Dakar; another was working in Senegal, at Koungheul. One of his younger brothers was working in Marseilles; another, with a son of the Dakar-based brother, was in Paris.

All four of the household's married women farmed that year, at Bema, Jaabalu, and Dogo. They harvested only a few *kande* each of immature groundnut pods.

Siliman Sanqare belonged to the Kuŋani farming group, as did all four of the household's married women. He held the positions of pump operator and storekeeper.

Jaabe So

Kuŋaninko t'i na makka xa soxono; the Kuŋani farming group decided to grow maize as well as rice, as we felt that maize would be more profitable. There was trouble with the pump at the start of the season. We grew a collective field of six acres of rice and six acres of maize and *feela* sorghum. We harvested about fifteen tons of grain, not counting the ears of maize sold fresh for roasting. We had a good harvest; but when SAED came to present its bill, ah! we had a hard time paying it. I went round to all the *groupements* to say they themselves should keep a written account of everything they took from SAED. There were too many *paysans nus* at that time. What I mean by 'naked peasants', is peasants who can't keep accounts.

Many of the groups had poor harvests of rice, because of sandy soils, weeds, pump breakdowns, and the lack of good advice. People were discouraged by the high costs and poor results of rice-growing.

Manayeli Jallo

Jaabe t'o na maxa tanbon wutu. Jaabe always said we shouldn't go into debt, we shouldn't get into the habit of receiving things without paying. But SAED kept on at him, until he finally said, 'It's not my decision alone; I'll call a meeting.'

Jaabe So

O su su, all of us, from Baalu to Gande, had agreed not to sign the SAED contract. I didn't know it at the time, but before we were due to meet at the *mairie*, Seyidu Ñaŋaane had contacted certain towns on behalf of SAED.

Well, we held our meeting, in October 1977. When we had gathered, I invited the vice-president to speak. Seyidu said: 'Now that we're all here, I can say that what we in Baalu want is to sign a five-year contract.' The Arundu delegate said that was what they wanted too. Mammadu Kande said the same thing: 'We've had enough of meetings, let's sign it and see.'

The president of Bakel, Saajo Fenda Tarawore, said: 'I don't understand this. We all agreed that we wouldn't sign the contract. Now you say we should sign it.' Waagi Sumaare, the president of Yelingara, said he didn't understand it either. The president of Tiyaabu, Sili Tapa Bacili, said: 'Seyidu, before you decided to sign the contract, did you discuss it with all of us? It's you that SAED has been visiting: Arundu, Baalu, they've been back and forth there.' The president of Yafera, Laaji Timmera, said: 'Seyidu sent a message to me, saying we should sign the contract. I said we should wait until the meeting, and all decide together.'

At that point, I said: 'Listen, Seyidu, I don't understand what you're up to. Arundu, Baalu, Gallaade, Mudeeri, I don't understand what you're up to. I can't keep you from signing, but I can resign.' Seyidu said: 'Well, we're going to sign it. Baalu, with all its members, Arundu, Mudeeri, Gallade: we've all decided to sign.' Manayeli Jallo said: 'Gande, Jawara, Manayeli, Tiyaabu, Bakel, Kuŋani, Yafera, we've all said we won't sign. Now you say you're going to sign. You're the ones creating trouble.'

A SAED agent asked to speak, and Seyidu told him to go ahead. He said: 'What I asked you to do was to reach an agreement in favour of signing. Just agree.' I said: 'Agree with whom?' That really made me angry: that certain towns had come to terms with SAED behind my back. 'Go ahead. But if you're going to sign the contract, I resign.'

I got up to leave. The presidents of Bakel, Tiyaabu, Manayeli, Yelingara, Yafera, and Mudeeri came after me and asked me to return. I went back into the room and sat down. They said: 'We won't continue with the meeting; the way things are today, we won't be able to agree. We left the meeting to go after our president. Now we'll have to see how we can reinstate him. But the meeting is over for today.' So everyone left.

Afterwards, in December, they called a meeting of all the farming groups. Mammadu Kande Gunjam of Gallaade was the first to speak. He said: 'We became involved in something we didn't understand. Seyidu came to see us to say we should sign the contract with SAED. We thought it was your idea as well. When Seyidu said we should sign

> The Director of SAED announced that the project would henceforth be governed by the following rules: respect for work-norms prescribed by SAED; gradual increase in irrigated surfaces farmed per village, to make full use of the pumps' capacities; individual plots for each family or household; signature of a contract between SAED and each groupement.... By 1981, a total of 1,780 hectares is to be under irrigation.... Kuŋani is to farm 100 hectares.
>> (Minutes of a SAED meeting held in Bakel in December 1977, attended by twenty-seven people: among them three Americans, including the Director of USAID in Dakar, three French employees of SAED, and two representatives of farmers' groups)

the contract, you should have said: "No, we're not going to sign it." You shouldn't have resigned. Who is there to take your place?'

Jawara, Yelingara, Manayeli, Tiyaabu all said they were with me from the beginning. The president of Tiyaabu said: 'Jaabe, the position is yours and no one else's. Please don't do that again. It was you who came to organize us all. You organized us, then came the fight with SAED. For you to resign and leave us with SAED, is a betrayal.'

I asked the assembly to forgive me. I said it had made me angry to realize what had been going on, SAED visiting Seyidu in Baalu and driving him around to talk to all the groups, all that plotting going on under my very eyes without my knowledge. That was why I said I'd resign. I acknowledged the truth of what they said, and took my place as president once more.

Adrian Adams

In December 1977 the Director of SAED, Cheikh Cissokho, held a meeting in Kuŋani. He announced the start of a new project which was to cover several thousand acres in the next few years. Kuŋani, he said, was to grow twenty-five hectares next year. All *groupements* were to sign a contract with SAED, and SAED had decided that land must be divided among individual families.

Lucien was part of the official delegation, along with several other Europeans no one in Kuŋani had ever seen before. He spoke to none of the townspeople.

In February 1978, the President of Senegal, Léopold Sédar Senghor, made a pre-electoral visit to Bakel. At the welcoming ceremony, Federation representatives had to struggle with *anciens combattants* and political groups to place a small table displaying their produce. During the afternoon, President and Madame Senghor were scheduled to visit the Bakel farming group's irrigated fields. When representatives of the Bakel group and the Federation went out there, they found that access to the fields was blocked by policemen who forbade them entry. The president of the Federation had to argue vehemently for the party to be allowed into the field to await their illustrious visitor. As the trip was running behind schedule, he spent only five minutes there, during which time he spoke only to two French technicians working for SAED, come down from Saint-Louis for the day. So the prepared speech could not be read out; but Jaabe So managed to thrust the written text into President Senghor's hands just before he left. It explained the Federation's conflict with SAED, and added:

For us, true development must be community-based . . . Agricultural development must first of all satisfy the needs of the people involved. . . . As it is, farming groups which have heeded SAED's advice have not done well out of it; having given up their rain-fed fields in order to grow irrigated rice, they have ended up without sorghum, and with a quantity of rice quite inadequate to feed their families, and which, sold to SAED at the current price, will not even enable them to pay what they owe for fuel, fertilizer and so on. So their efforts will not have earned them anything. That is not true development.

Jaabe So

During the dry season, Kuŋani grew vegetables; but we grew less than before, as did other villages, because of marketing difficulties.

Robert's replacement as *chef de périmètre* in Bakel, Niang, visited Kuŋani with a group of local government officials. He repeated what Gawusu had said at the earlier meeting, and read out the SAED contract. There were several points we did not agree with: they said collective fields would not be allowed, and the farming groups would have to set up a joint bank account with SAED. There was heated discussion.

I told Niang that it was one thing to make people responsible. Two years earlier, I'd already suggested to Robert Laborde that each work-team should be responsible for a specific part of the collective field; that was my own idea, not SAED's. But wanting to make everything individual, with everyone just farming his own household plot of land, was a completely different idea. It wouldn't work with irrigated farm-

'It all began a few years ago when this fellow from the area went to France to work and was impressed by the technology he saw there. So when he got back to the area here he wrote to the Canadian agency in France and they sent out a French technician to help. After then the English war on hunger organization came in and also aided. Two years ago the people had formed their own organizations and were doing quite well, but did not have all the necessary technology and financing, so they approached AID, and the project came about....

'The people hit SAED with the proposition that it provide them with heavier equipment. The project has merit because the local people themselves will be running the darn thing.'

> (Remarks by the Director of USAID cereals projects in Senegal, Dakar, June 1978; noted by Richard Franke and Barbara Chasin)

After five years in the field, after four years of work both on the improvement of traditional agriculture and the introduction of a new form of agriculture, we believe that we have data which could be of use to people responsible for the reinsertion of emigrants. From these data, first based on an empirical discovery of the Bakel region, but then based on constant research, an in-depth socio-economic investigation and constant re-evaluation of our activities, we shall now attempt to derive an approach to a possible 'model' for émigrés establishing themselves as farmers.... For our study, we will take the case of two emigrants and their families (10 people in all) working together to set up a modern farming concern. [The document goes on to explore two possibilities: one for irrigated farming by the river, with a motor-pump and animal traction, based on five hectares of rice and two hectares of rain-fed sorghum during the rainy season, five hectares of irrigated maize during the dry season—total investment 3,355,000 CFA francs, about £8,000; one for rain-fed farming with animal traction, based on eight hectares of sorghum, four of maize, two of groundnuts—total investment 1,420,000 CFA francs, about £3,350.] ...

Does development require the dismantling of traditional structures? The example of the village of Somankidi in Mali [where 14 emigrants of 4 nationalities have set up an irrigated perimeter] seems to confirm the views of those who favour that thesis. But the results in Matam, with small village perimeters organized by SAED, give cause for hope that a degree of efficiency may be compatible with respect for structures...

> (From a paper entitled 'Reinsertion of African Emigrants in Modern Agriculture', sent to Jaabe So by Robert Aprin in June 1978)

ing; there were too many risks. Besides, dividing up plots among household heads would mean there was no place for women.

I also disagreed with SAED's idea that people should grow only rice. I asked Niang: 'What will you pay us for a kilo of rice? Forty francs? That's four thousand francs a hundred-kilo sack. The cheapest sorghum costs more than that; and we live on sorghum and maize here, not rice. Rice is in your interest, not ours. If the government told you to take pumps away from people who refuse to divide up their collective field, who refuse to grow only rice—then take your pump and go.'

I wrote to the Director of SAED to explain these things. I also said that farmers should be kept informed about plans for developing the River.

Adrian Adams

The clause in the SAED contract requiring individual family plots was modified to permit a maximum of 30 per cent of irrigated surfaces to be farmed on a collective basis: the demand for direct SAED control over *groupement* finances was softened to a demand for permanent access to information.

A French geographer, under contract to War on Want to do research on their project on the north bank of the river, visited Kuŋani briefly in June 1978. The project's aim was to improve traditional agriculture. There was a French agronomist working on it, the third agronomist since the project started. A book on the traditional agriculture of the area was being produced.

She also mentioned that there seemed little prospect of establishing a development project on the north bank, because landowning families would not release land for that purpose, and there was no legal means of dispossessing them. Perhaps something might be done in the nearby administrative centre, where the government owned the land. Also, the Mauritanian equivalent of SAED, SONADER (Société Nationale de Développement Rural), was considering setting up irrigated perimeters in the area.

7

Epitaphs
(1978–1981)

Year Six: 1978/1979

Jaabe So

Kuŋaninko da te xooren dabari; the Kuŋani farming group again planted a twelve-acre collective field, half rice, half maize. Also, for the first time, twenty-five acres were divided up among individual members. Each man was given two plots, each one twelve metres by twenty-five, and each woman was given one plot for growing rice, with two women sharing a plot for growing maize and okra. Everyone felt this was fair, because men farm to feed the entire household, whereas what women harvest is their own.

Ever since the visit to Matam, I knew that if SAED's system of dividing up the land among household heads were applied here, women would leave irrigated farming. The women here are independent farmers; in the fields, they don't work for men. They need to have land in their own right, and keep or sell the produce as they choose.

There was a SAED *encadreur* who visited us from time to time. He was the third we'd had since the French technicians left. The first one said SAED did not provide good working conditions; the second said he hadn't been given proper training. The third said SAED's attitude made it difficult for him to have decent relations with the *groupements*.

The only farming groups that didn't grow maize were those who hadn't the right kind of soil. Bakel, Manayeli, Yelingara, and Gallaade grew only maize. Arundu, Tiyaabu, Jawara, Gande grew half maize, half rice. The rains were not good that year. In the dry season, the Kuŋani farming group did only a little market gardening.

I didn't farm Seegankaani that year. For the next four years, I didn't farm at all; my household lived off my savings, and when I travelled, or people came to see me, I paid the expenses myself. SAED had cars and money, they could go everywhere, hold meetings and say what they liked. The only means I had was what I'd saved from when I had a shop in Kuŋani and baked bread; even that wasn't very much, because during the drought when people were hungry, I gave people food from my shop on credit, and many of them never paid me. I wasn't yet eligible for a pension from the merchant marine.

One day my wife Maamu asked me: 'Do you know how much you've spent?' I said: 'Yes, I know.' She said: 'Can things continue like this? You'll wear yourself out, and have nothing left.' Well, she was right; but what could I do?

Mammadu Lasana Ba

I ti soxaano na maaron soxo; SAED said the farmers should grow rice. But rice can't be grown without a properly levelled perimeter. And some soil is good for sorghum or maize, not for rice.

Manayeli Jallo

Kontran falle, after the contract was signed, Jaabe said they shouldn't force us to grow rice. There were some towns whose fields weren't good for rice; we should refuse. We didn't listen at the time.

In Manayeli we grew twelve acres of rice. I was the first pump operator. We kept the pump working the whole day long; a plot could be knee-deep in water, but an hour later it would all have disappeared. The rice grew only knee-high. It was no good. We owed a lot of money to SAED; our *caisse* was weakened.

At that time, the *groupements* had money. Some people said that Jawara harvested 150 sacks of sorghum from their collective field. It all went on paying debts to SAED.

Jaabe So

Sayedin ga da mesini riti, when SAED brought pumps, they pushed everyone into growing rice. How can rice do well on sandy soil? People kept irrigating, over and over again; they spent more on fuel than the rice crop could ever be worth. Finally they decided to grow sorghum or maize instead. I myself told them: 'Anywhere that's not floodland or land that holds water well, if SAED tells you to grow rice, just say no; grow maize or *feela* instead.'

When Yelingara stopped growing rice, SAED went and took their pump away from them. There was trouble; we fought SAED over that, until they gave the pump back again. When Tiyaabu grew rice, it dried up before it could grow tall, and never bore grain; they irrigated over and over again, but it was no use. Was SAED paying them? Had SAED hired them as labourers? All the farming groups agreed to grow maize or *feela* wherever the land was not good for rice, whether SAED liked it or not.

By then, we had no technician any more. We were our own technicians. The technicians we had were no use to us. They would even have done harm if we hadn't been awake. Even those who said they agreed with us didn't help us in the end; they just sent researchers. And we realized that researchers aren't looking for the truth. There you are.

Adrian Adams

In April 1979 Jaabe So's son Mammadu, aged twenty-one, went to Dakar to begin his apprenticeship in motor mechanics.

Manayeli Jallo

Jaabe ga da golle ke riti, when Jaabe brought the work he didn't just build the room, he furnished it. He knew that we couldn't manage unless we had a mechanic. He didn't start by saying, 'Let's send my son to train as a mechanic.' He asked Matforce in Dakar to help him train five young men as mechanics. He told all twelve *groupements* about this, from Baalu to Gande. We met and discussed it. Jaabe's son wasn't named that day! Yelingara proposed someone, as did Gallaade, Gande, Tiyaabu, and Baalu. Jaabe noted their names and sent them to Matforce.

When the time came for them to go to Dakar, it was just before the rains, and they all said: 'My son can't go, I need him for farming.' Jaabe was ashamed not to send anyone after making all the arrangements; he said, 'You've really let me down.' So he sent his son Mammadu, the only one of his children who was of an age to help him. Mammadu didn't want to go; he was planning to go and work in France. Jaabe made him go to Dakar instead.

Adrian Adams

By chance, I came across the first draft of a 'Socio-Economic Study' written for OMVS. Having read the section on Bakel, I went to see the director of the team of OMVS experts responsible for the study, and told him that it contained errors of fact and interpretation. He suggested that I help rewrite it.

In June 1979 I went to OMVS headquarters in Saint-Louis with fifty-odd pages on the Bakel area, divided into two parts as required by

Despite certain individualistic tendencies encouraged by the scale of migratory processes and the ensuing high inflows of revenue, communal traditions remain very strong among the Soninke. Is this why irrigated farming was introduced in the name of a community-based ideology? It seems, indeed, that the initiator of this experiment, who shortly after the first farming season became 'president of the Federation of Bakel area village work-groups', often referred to this tradition ... There were many obstacles in the way of a collectivist undertaking of this type, seeking to transcend individual experiments and impose the village as a unit of production. Among others, there is the lack of unanimity among villagers and the lack of fit between the new hierarchy created in the 'farming group' and the traditional social structure ... No doubt it is with reference to the personality of the president of the Federation, himself a man of artisan caste, that a researcher who has closely followed this case pinpoints this as one of the weaknesses of the proposed system. ...

It is beyond doubt that the recuperation of the experiment by SAED, after it had been in operation for a few inconclusive months, was opposed neither by regional political personalities, nor by the traditional authorities ... Does not one particular man or group of men bear responsibility for this partial or total failure? Is not the non-re-election, in early 1978, of the promoter of the experiment as president of the Federation of village farming groups, the price paid for this failure, acknowledged as such by all responsible elements of the population of the region?

(OMVS, 'Socio-Economic Study', First Interim Version)

the study's general plan, 'Traditional rural environment' and 'Introduction of irrigated farming'. After extensive discussion, and minor changes to the text, it was accepted. The passages to which I had specifically objected were replaced by an explanation of current differences between SAED and the Federation, which concluded:

The ground lost can still be regained. The existence of the *groupements* and their Federation, brings out into the open contradictions often glossed over by official reports, but which can still be overcome: between self-sufficiency and the market, between subsistence crops and cash crops, between peasant self-management and management by bureaucracy; between the rhetoric of proclaimed priorities (subsistence needs of River people, their need for higher and more secure incomes) and the reality of an approach based on extending intensive rice-growing as quickly as possible (with consequent pressure on farmers to neglect subsistence farming, and to sell most of their irrigated crop to pay the debts incurred).

Epitaphs (1978–1981) 161

In October 1979 I found that whereas the first part of what I wrote on the Bakel area, dealing with 'the traditional rural environment', had been incorporated in the second interim version of the OMVS 'Socio-Economic Study', the second part, on 'the introduction of irrigated farming', had not been included. Instead, there was a brief summary written by someone else; and the part of the report which had prompted my initial protest was reinstated in its entirety. When I asked the director for an explanation, he said that my text had not been included because it represented the point of view of peasant farmers, whereas the study was intended for technocrats. I then sent him a revised version of the summary of my initial text on 'the introduction of irrigated farming', excluding the part of the report I had objected to; stating clearly that nothing less than that would be acceptable.

In 1980 a visiting OMVS expert gave me a copy of the third interim version of OMVS's 'Socio-Economic Study'. It contained my revised version of the section on the introduction of irrigated farming in the Bakel area in the second interim version of the Study; that having been a summary, written by someone else without my knowledge or consent, of my own text on the same subject, which I was invited to write after I objected to the text published in the first interim version of the Study.

Finding thereafter that visiting experts, including those hired by OMVS, continued to ask questions revealing a profound ignorance both of traditional agriculture and of the way irrigated farming was introduced to the Bakel area, I began asking them whether they had read the OMVS 'Socio-Economic Study'. They all said no. One of them laughingly added: 'No one reads that kind of stuff; just the conclusions, if that . . .'.

Jaabe So

Guwerneerin d'in xiri Tanba; in July 1979 the Governor summoned me to a meeting in Tambacounda. Five Federation people went, along with the Bakel *préfet* and Niang, the SAED *chef de périmètre* in Bakel. The problem was that SAED wanted us to sign a new contract, for five years. All of us peasant farmers knew they hadn't respected the two-year contract they'd made us sign before, whereas we'd respected our share. So the whole Federation refused to sign the five-year contract. Niang informed Saint-Louis, Saint-Louis informed Tambacounda, and they summoned the peasant farmers to a meeting.

We spent the night in Tambacounda; the meeting took place in the

In the Bakel area, unlike others, irrigation began as a result of local peasant farmers' initiative. Wishing to improve their farming techniques to counter the effects of the drought, they obtained the assistance of the Compagnie Internationale pour le Développement Rural (CIDR), which trained technical personnel for small agricultural projects funded by agencies aiding the Third World. In 1973 CIDR signed an agreement with the Senegalese government, according to which it was to help develop the Bakel area and reinsert returning migrant workers.

In 1974–5, after the arrival of a CIDR technician funded by War on Want, a farming group was created in the village from which the initial request had come. An area of fonde land surrendered by its owners was farmed collectively, growing maize and sorghum during the rainy season and vegetables during the dry season. Three other villages of the Bakel area began growing vegetables during that first dry season. Faced with this initiative, the authorities decided in 1975, after a visit by the Minister of Rural Development, to develop the Bakel area according to the methods evolved in Matam under SAED control.

The leaders of the first village groupement, fearing that peasant farmers might lose control of the movement they had created, had a number of objections which they clearly expressed during a meeting with SAED in March 1975. When the CIDR technicians were co-opted by SAED, that struck them as proof that their needs were not being considered; a few months later, the Bakel area farming groups created a Federation in order to defend their interests.

The Federation defined its status, its working principles, and its view of the co-operation it expected from SAED, according to which SAED's contribution would be mainly technical. It also outlined the kind of development it considered appropriate to the area: complementary association of irrigated and rain-fed farming, priority to subsistence farming, and initially at least, collective organization of irrigated farming, in order to minimize its risks in an area where manpower is in short supply. These suggestions were distinctly at variance with SAED's style of management as stated for instance in the text of their 'Small Perimeters Contract'; SAED, of course, merely executing Senegal's national development policy. . . .

Despite certain individualistic tendencies encouraged by migratory trends and the relatively high incomes they entail, the communal tradition remains strong among the Soninke. Is that why irrigated farming was first introduced in the name of a communal ideal? Its initiator, who subsequently became president of the Federation, often referred to the tradition of communal labour. But there was in fact a considerable degree of innovation, as the formula put forward by the president of the Federation was collective farming at the level of the community as a *whole, with not just a family field, but a community field, being farmed by a groupement made up of all those who wished to take part, each person being a full member without distinction of sex or age.*

There were many obstacles to setting up such a system based on communal principles. Some were internal; such as conflict between villagers (organizing collective work requires a degree of consciousness and self-discipline rarely to be found at the outset), and conflicts due to the persistence of inegalitarian attitudes (as between young and old, or in cases where an incompetent leader, chosen by virtue of his social position, blocked initiative and progress). It must be noted, however, that such problems, which are not peculiar to the Bakel area and indeed in some respects are less acute there than elsewhere, were intensified by the conflict between peasant groupements which on their own initiative took on the risks and responsibility of introducing agricultural modernization to their area, and a supervising agency with a bureaucratic and urban image, sending out from Saint-Louis technicians who were sometimes incompetent and ill-informed, and tried to impose their choices and decisions on those peasant groupements.

However, there are signs to be observed in the area which give reason to hope. It has been noted that the current problems of irrigated farming in the Bakel area, arise on the one hand from the supervising agency's lack of understanding of local environmental and social factors, and on the other hand from the peasant farmers' lack of understanding of the constraints and possibilities of the new techniques being placed at their disposal, the two being linked. One can hope that the present report will help remove the former obstacle, and thus, in time, the latter.

(OMVS, 'Socio-Economic Study', third interim version, 1980)

---

morning. There were many people there; another meeting was to take place that day, and the Governor said he'd hold ours first. All the region's *députés* were there. The Director of SAED spoke right away, saying that the peasants were disobedient: we didn't listen to them, we were in debt, we didn't want to sign the contract. Then the Governor spoke: he said we should respect SAED and pay our debts.

It was the Bakel *préfet*'s turn to speak, but he said they should let me speak first. I said to the Governor: 'I was pleased to come to this meeting, because I thought there would be justice here, for us and for SAED. Now I see that your justice is no good. We respect SAED; does SAED respect us? Peasants must be respected too; we must be asked what we think. Governor, I must tell you that peasants are not slaves.'

I raised the problem of debt. I said: 'The Director of SAED has mixed up debts that belong to different categories. Our Federation has paid fourteen million francs, with 180,000 francs left owing. He's mixed that up with the debts owed by the *groupements* on the *Fanlenme* to make trouble, because we haven't signed their contract. We know how much we owe. It's 180,000 francs. Is that clear?'

The whole *salon* began to demonstrate, with everyone shouting and clapping. The Governor said: 'Stop, stop!' People fell silent; I began to speak again. I said: 'Governor, what use is SAED to us? All it has brought is the pumps on the river. They put pressure on people to bring more and more land under irrigation, with no technicians to help them. We don't respect SAED any more, because we've seen that its *encadrement* is very, very negative. Before SAED existed, our country lived by farming. Whoever wants to supervise us has to know the right way to work with us. It's not enough just to consider your own interests; you have to consider the peasants' interests as well. Someone who only considers his figures will never get along with us. But you, you assume straight away that it's SAED who's in the right.'

The Governor said: 'I hadn't understood it like that, that's not how it was explained to me.' I said: 'No, that's not how it was explained to you, because the Director of SAED has lots of money, he can come and spend two or three nights here. But we also count as Senegalese citizens. The Director of SAED should speak to us. If he doesn't respect us—we're not dogs, you know! A man should be considered a man.'

The *préfet* of Bakel spoke next. He said that SAED had apparently not respected the first contract. This time, he would keep an eye on things himself. Then the *député* of Tambacounda spoke. He said to the Governor: 'You should not treat these peasants as ignorant, and expect them to work for you while you disregard their interests.'

The Governor said everyone should come to his place to eat. I was sitting in a corner, with my food and a bottle of *limonade*, when the Governor came up to me and said: '*Monsieur* So, we ought to get along together. I'll come and hold a meeting in Bakel, and we'll come to an agreement.' I said: 'We'll come to an agreement if a just decision is made. As far as I'm concerned, we could have come to an agreement right here. But you brought us here to frighten us, to deprive us of our rights. In Bakel, if things aren't judged rightly . . . I'm free, free to give up this business and go and sit quietly at home.' He said: 'No, no, we'll come to an agreement.'

We came home. On the third day, the *préfet* summoned me. He said: 'So, you're right; but I'll keep an eye on things. Sign the contract.' I said: 'You've been fair to us; we'll sign for your sake—for two years. But mark my words: they won't respect it.'

Well, we signed a two-year contract. Nothing was respected on their side. I went to see the *préfet*. I told him, 'See? They've not respected anything. When a pump breaks down during the farming season, there's a twenty-day wait before it's repaired. When fuel runs out, we have to buy fuel from the Société Nationale de Distribution.' He said: 'It's true they've not respected the contract. We'll see about it next time.' The *préfet* was posted elsewhere, because he had spoken in our favour; his name was Oumar Cissé. And that was the last time we were asked to sign a contract.

Year Seven: 1979/1980

Jaabe So

O waqile, we decided to grow twelve acres of rice and as much maize as possible. Once more the work was delayed, and by the time canals had been repaired and fields tilled, people sowed their rice-fields hastily, in order to be free to sow their *jeeri* fields of sorghum and groundnuts as soon as it rained.

Once more there were problems with weeding the rice. Just when work was at its heaviest, three collective work-days a week had to be spent weeding, and sometimes an extra day as well, in addition to the days worked by each team in turn. We managed to save the rice crop. But this delayed sowing the collective maize-field, so that it was much smaller than intended; we put some of it off until the dry season. Work in the *jeeri* fields also suffered. By the time group members managed to sow their individual irrigated plots, the rainy season was half over. Each woman had two plots, and each man three; they did well in the end.

Adrian Adams

Towards the end of the dry season, the Federation deposited an application for official recognition at the office of the Governor of the Region of Sénégal-Oriental in Tambacounda: two copies of the association's statutes, four copies of the minutes of the constituent general assembly, four copies of the list of founding members, and a handwritten letter of application to the Minister of the Interior. A first application, drawn up in 1976, had been rejected as somehow not conforming with legal requirements, and there were great difficulties in ascertaining the procedure to be followed. Jaabe So had been told of a lawyer who might help, and wrote to him in October 1979:

> The farming groups belonging to the Federation deplore the fact that they have still received no acknowledgement of their application. They have authorized me to retain the services of a lawyer to enquire into the matter. I should be grateful if you could do that for us, on whatever your usual terms are.

Jaabe So

It was my nephew Bokar Sisoxo who put me in touch with Maître Babacar Niang. He didn't tell me he was in politics; he just said he was a lawyer who could help us with our application for official recognition. When I went to see him in Dakar, he took me to see a Doctor Diallo; they promised to help us, and I returned to Kuŋani. Then they sent for me, and I went back to Dakar, with Modibo Tanjigoora. We met with them at Doctor Diallo's house; they said they would help us. Then they brought out two cartons full of membership cards for a *syndicat paysan*; they asked me to take them home and distribute them.

I brought the cartons back here with me. But I told the others: 'This is politics; we mustn't get involved. We asked for help from a lawyer; what does that have to do with these membership cards? If we aren't careful, these people will do us harm.' I destroyed all the cards, and took our papers back from Maître Niang.

SAED normally avoided visiting Kuŋani; but that year they had to bring their visitors from the American Embassy to see us, because the vegetable seeds they had given other groups were no good. We'd bought our own seed, and ours was the only market garden in the area. After that, Gawusu himself brought some other Americans to see our dry-season maize fields, because they were better than anyone else's. They visited the collective maize-field; they congratulated us, and asked who taught us to plant maize that way. I said: 'No one; we learned it ourselves.' Then they asked: 'Whose field is that over there?' I said it was mine. An American took a stick and measured the distance between plants. He said: 'This field was sown by the same person who sowed the collective field.' I said: 'That's true; I showed them how to do it.' He said: 'That's exactly how it should be done.'

Our tomatoes began to bear fruit; then the plants suddenly withered and died within a few weeks. SAED said it was because we gave them too much water. We didn't believe that. We harvested two tons of onions, but while they were drying under shelter there was a freak shower of rain; many onions were spoilt.

We paid what we owed SAED in cash, partly with money from our *caisse*, and partly, for the first time, with contributions in cash from group members, 1,250 CFA francs per plot. That meant that each woman paid 2,500 francs, and each man 3,750 francs.

We didn't sell our paddy to SAED, for the same reason that we paid our debts in cash, not in kind: because SAED's price, 41 francs a kilo, was too low. We sold it in the town, at 50 francs a kilo. It sold slowly, being unhulled; less than half of it had been sold by the end of the dry season. The group also sold cabbages, tomatoes, onions, and maize. That year we earned about 860,000 francs. After paying our debts, our profit was only 150,000 francs. Each person also had the produce of his individual plots: say about 150 kilos from each plot of rice, and 100 kilos, dry weight, from each plot of maize.

By then, Federation farming groups as a whole owed SAED over seven million CFA francs. They managed to pay, but it took a long time. Some of them asked members working in France to help.

We wrote to the Minister of the Interior to ask about the Federation's statutes. We also sent a report on the farming year to the Minister of Rural Development, the Director of SAED, and the *préfet* of Bakel. There was no reply.

Al-Haji Mpali Tanjigoora

Many people left the group. The people who stayed on were those who could endure hardship and disappointment. Those of us who stayed understood that you can't have irrigation without spending money. Before, there were many people in the group, but they didn't understand that; they were used to *jeeri* farming.

The main problem, though, was that our fields were never levelled; when you irrigated, there were many places the water couldn't reach. If it hadn't been for that, many of those who left would have stayed on.

Adrian Adams

This was a time of great isolation. SAED's fleet of yellow Ford Cherokees came and went in the distance. There was no news of what might be happening in nearby Bakel, let alone Dakar, five hundred miles away. Chance unofficial visitors would hint at ominous things. Such local officials as one had to deal with, were hostile.

There is also a responsibility with respect to the ultimate recipients of the vast amount of development aid currently being mobilized in areas such as the Sahel. It is, hence, a responsibility to the very individuals who benefit from that aid. Based on the analysis, the findings and the data developed by social scientists—planners and politicians—local administrators and headmen and the families in thousands of villages, all will be impacted upon in a way which was thought impossible only a decade ago. Social scientists, therefore, have a unique opportunity as well as a profound duty if the village is to continue as the primary arena for human existence and growth in the Sahel.

(David Shear (USAID Director in Senegal), 'The Role of the village in Sahelian Development', in P. Reining and B. Lenkard (eds.), Village Viability in Contemporary Society, *1980*)

A three man evaluation team undertook a field evaluation of the Bakel Integrated Crop Production Project from March 28 to April 21 ... The team was in Dakar March 28 to April 4, researching existing documents and interviewing Senegalese and USAID project officials. From April 4–11 they did their evaluation in Bakel visiting most of the perimeters with GOS and USAID project officials. From April 11 to April 21 the team completed its report. ...

Since the beginning of the project, the roles of both farmers and SAED in the Bakel region have evolved as a result of pressures for increased management autonomy, production, debt management and marketing by farmers groups. ...

Findings

1. In principle, the Bakel Project is conceptually sound. It is focused on the key constraint, water availability, that has impeded improvement in the standard of living of the people in the area and has reduced agricultural production.

2. The role of SAED as supplier of inputs as well as provider of technical assistance is a necessary one at this stage of development, SAED's stated policy in Bakel of passing to Project participants more of the responsibility for management, as private institutions develop, is supported. ...

3. The relatively low number of hectares under irrigation at this time should not be interpreted as failure. It does signal serious problems in implementation. However, the past few years have afforded opportunity to SAED and the villages to experience the difficulties in trying to introduce irrigated agriculture in the area. ...

5. The experimental nature of the Project should be recognized. ... The benefits from the project will come not from the few hectares brought under irrigation during the life of this project, but rather from

replication of a refined system of irrigated agriculture in later years. The present state of the art, as practiced in Bakel, with high costs and low yields allows insufficient margin for error and runs the risk of being rejected by the producers unless efficiencies are improved . . .

7. The life of the project should be extended for another three years to 1984. . . .

Sociological Aspects

. . . On various perimeters the farmers looked upon the success of irrigation farming with mixed feelings: Some farmers were reported to have left their groups; some groups were smaller than when formed; some farmers were fearful of going further into debt; others had decided that their time was better spent working on their rainfed land. However, the average size of each perimeter has grown over the years. . . .

Institutional Analysis

. . . The Federation of participating groups, centered in Kounghani, has played a continuing role during the life of the project. The original contact by foreign technicians was facilitated by the federation; only later was SAED brought in as the principal institution for introduction of water management. Although not formally assigned this role, the Federation has acted, in effect, as a major negotiator concerning contracts, purchase price of rice, costs of inputs, and modes of debt management. . . .

SAED Contribution

. . . The Project staff reports that the work performed by the technical divisions of SAED is of extremely poor quality and in many cases unusable. Perimeters were constructed with faulty information on soils and slopes and water delivery systems are correspondingly inefficient.

(Final Report, Joint Assessment of US Assistance Program in Senegal, 1980)

We received no reply to a note written at the request of a visitor from USAID in Dakar.

How can the Federation hope for US assistance, when everything we have experienced so far suggests that USAID works only with Senegalese government agents? We fear that any suggestions we might make, may lead only to greater US support for SAED. In that case, the only thing you can ask for on our behalf, is that the US cease all aid to the Bakel area. USAID's 'Bakel project' has done the organized peasant farmers only harm. It was AID money that attracted a large SAED team to the area; they took away from us the Oxfam/War on Want-financed technician we had working with us, even the

A more discreet start, and a desire for autonomy translated into efforts to raise as much funding as possible from within the community, locally and from migrant workers, would have made it possible to prolong work at grass-roots level.... That might have made it possible for the project to evolve differently. More precisely, we are thinking of the high level of motivation and commitment shown by the population during the year of 'free work', that yielded to passivity as the State apparatus took over. A better understanding of Sahelian Africa, of its position within Africa–Europe market mechanisms, of the strengths and weaknesses of the population, in 3 words of the 'assets and liabilities' of the project of the Bakel Soninke, would have induced the technicians to be more cautious (restricting their outside contacts, warning of the risks inherent in motorization), and possibly to decide to do without motorization altogether. Failing that, visits from advisers, and a close coordination of people and organizations willing to help the Soninke, might have made up for certain deficiencies. But such cooperation did not take place.... Once USAID's 'gifts' and SAED's 'assistance' had been accepted... there remained only for the Soninke to seek an acceptable compromise....

At its birth, over four years ago, the Federation was surrounded by partners: USAID which opted for neutrality, SAED which was more or less hostile but prepared to come to terms, the CIDR technicians and War on Want who were wholly in favour. Today, the Federation has practically only one interlocutor, SAED, since USAID has restricted itself to a funding role, and the other organizations have left the project. It can therefore rely only on its own resources, which is no bad thing in itself; but loyal allies might have been useful at times....

Certainly, the CIDR technicians made mistakes. It is understandable that War on Want should not have wanted to fund a project along with USAID, whose goals and ideology it does not share. It is commendable that as a result of in-depth socio-political analysis, War on Want representatives should have been sceptical about motorization and irrigation. But War on Want should not have 'dropped' the Soninke peasant farmers at the time when they most needed support, after having led them to hope otherwise up to the last minute....

The 'Fédération des paysans organisés en zone soninké de Bakel', born of the Soninke resistance of the 'spring of '75', does not fundamentally call into question the capitalist system, but it shows that there is another way of looking at rural development: a 'développement paysan' instead of a 'développement technocrate et fonctionnaire'. If it can overcome its internal problems, and avoid the trap of perpetual assistance, the Federation will contribute a great deal to the valley's development. Most of all, it will spare the peasant community the proletarization which awaits

them once the valley is developed. That proletarization is inevitable if a capitalist, agro-industrial type of development takes place....
The strength of the Federation, and of the area's peasants as a whole, will depend on its ability to set off a popular movement throughout the valley. It is too early to know whether the Federation will be able to do so.... Whether the Federation will achieve its goals, will depend on the help it receives, and especially on the will of its members, who will need to rely more and more on their own resources. That Maoist idea should be adopted by all developing countries: it represents the only way towards their people's liberation.

(Robert Aprin's thesis, 'Development and Peasant Resistance: The Case of the Soninke of Bakel, Senegal', 1980)

A confidential report submitted to OMVS in 1979, reveals what far-reaching effects the Senegal River development plans will have on the peasant masses....
Increasing food production, freeing peasant farmers from the vagaries of climate, enabling them to develop their environment according to their wishes, and providing an alternative to badly-paid work in the cities: all that would have been possible if a project had been conceived in cooperation with grassroots movements, based on the gradual development of appropriate technologies remaining under community control. That would have been possible with much smaller investments, on the basis of communal responsibility for local and regional development. But such an alternative does not meet the needs of investors, who demand that production be geared to the external market.

(Report by a Canadian study group, 1980, given to the president of the Federation by a visitor)

---

vehicle we had been given; we have been denied recognition by the Senegalese government because they fear this might weaken their unquestioned hold on money made available for the Bakel area; and we have not been able to get even modest help from other sources, because our villages have been labelled part of the 'Bakel project'. We cannot stress too strongly that this damage has not been counter-balanced by any benefits.

A young French researcher, a student of Georges Balandier, visited Kuŋani. He was preparing a thesis on Senegal River development; he'd wanted to write on agrarian capitalism, but decided it would be premature. According to him, River development necessarily involved splitting extended families into nuclear families, and displacing the population according to the manpower needs of large irrigation schemes; 'since it's the people with money who decide'. People like Mamadou Dia, he remarked, wanted to base rural development on existing communities and values. 'But of course, that didn't make any sense. There's no such thing. *Les communautés réelles, ça n'existe pas.*' They'd told him clearly at OMVS: 'We've no need of the lineages.' Organizations based on tradition, like the Federation, had no future.

Year Eight: 1980/1981

Jaabe So

Yugun do yaxaru, there were now 136 people in the Kuŋani farming group: twenty-eight men and 108 women. We wanted to grow a six-acre collective rice-field, transplanted before the rains came. We repaired the irrigation canals and built new ones, in order to have more land for individual plots. We did all that by hand.

Rapairing canals was hard work, and the women felt they did more than their share. They said that collective work should be shared out among group members: so that, for instance, each woman would repair ten metres of canal a day, and each man fifteen metres. When I asked them why, they said: 'Some women don't come until noon; some don't come at all, even though they're not cooking for their household that day. We want women who stay away without an excuse to repair twenty metres instead of ten when they come the next day. Whenever they don't do their share, they should pay a 500-franc fine.' So that was how we did things from then on.

We irrigated six acres before tilling it by hand. We had decided to try 'Dapog' seedbeds: you soak paddy in water overnight, then spread it on a thin layer of soil over plastic sheeting. But the wind that blew from the north was so hot that it dried up the seedbeds; in spite of all the watering, we had to sow them again. Some SAED people came to visit us; the local man started telling us off, saying what we'd done was no good. A man from Saint-Louis told him: 'Just keep quiet; you may never have heard of it, but it's a perfectly valid technique.' They took some photographs. Next year we heard they were showing people in Baalu how to do it.

We gave it up later, because the women said planting out the seedlings was too much work at a time when they also had *jeeri* farming to do. We found something else that worked just as well: we irrigated, then trampled the mud and sowed rice broadcast, rice that had been soaked in water for twenty-four hours, then left in sacks for twenty-four hours to begin to sprout. That gave the rice a head start on the weeds. If only our fields were properly levelled, we would do that all the time.

It rained, and the *jeeri* fields had to be sown at once. The rice was planted out afterwards. By then it was too late to sow a large surface with maize as we had hoped to do. Late in the season each man sowed three plots for himself, and each woman two. Each man harvested about 300 kilos of maize. Each woman harvested about 150 kilos from her one plot of rice, and earned about 5,000 francs from the sale of fresh ears of maize, also harvesting about sixty kilos of dry maize.

The rains were not good that year.

Adrian Adams

Upon arriving in Bakel, the chief USAID technician Khoi Lê visited Kuŋani to announce his intention of working with the Federation, saying he hoped to do for the peasants of Senegal, that which he was not able to do for the peasants of his native South Vietnam. He then was not seen again for eight months. Upon reappearing, he explained that SAED had not wanted USAID to work with the Federation; but now there was a new Director of SAED, with quite different attitudes, and all would be well. This was to be the pattern for his subsequent dealings with the Federation.

When I was in Dakar in September 1980, Khoi Lê asked me to give a talk about the Federation to a group of USAID personnel. I accepted reluctantly, not wishing to seem uncivil, nor to be blamed afterwards for having refused to speak to them. The occasion was unmemorable, and I soon forgot it. But some twelve years later, I came across a document that placed it in an unexpected light.

Jaabe So

Ken su, all this time we kept trying to find out what had happened to the application we had made over a year before, to have the Federation officially recognized. Finally a high-placed official, Ousmane Goundiam, whose family is from Kuŋani, told me that the application was being blocked for 'political reasons'.

We wrote to the Ministry of Rural Development, in a letter that received no answer:

We have been given to understand that our application has been blocked because we have been classified as an 'anti-government' organization. We reject with indignation all such suggestions. We have no political activities nor

Over the years, one of the key issues for the Federation has been the lack of official recognition by the government of Senegal. Ongoing contacts between either the president of the Federation, or his wife (of British origin), and members of the SAED and USAID staff continue to raise a multiplicity of issues centered around the problems of local autonomy and control. A presentation by the Federation president's wife at a USAID brown-bag lunch September 29, 1980, in an air-conditioned Dakar conference room was instructive in this regard.

Prior to this meeting, many differences had arisen between USAID, SAED, and the Federation. The SAED director in Bakel had been replaced a number of times, each one having a different response to the Federation. Initially, the director of SAED preferred to ignore the Federation and its demands. This was followed by an attempt to defuse the Federation by introducing ethnic rivalry into the situation. The director of SAED at that time asked the Tukulor villages in the southern part of the Bakel zone to form their own Federation to compete with the Soninke Federation in the northern part of the Zone. However, the Tukulor Federation never became a viable organization.

Despite difficulties with SAED, the local USAID technicians had begun to work more directly with the officers and members of the Federation. The local USAID personnel supported the basic precepts of the Federation and made a sincere effort to provide technical assistance for those working on their irrigation schemes.

Still the nagging problem of official recognition resurfaced. In the September meeting, the USAID audience was reminded that SAED was still the dominant force in the Bakel zone. . . . It was argued that since both AID and the Federation were increasing their level of cooperation, it would be in their mutual interest for the Federation to achieve this goal.

The president's wife then offered a brief review of the process by which the Federation had sought recognition. The president was approached, she related, through a relative of his by a group of businessmen in Dakar who expressed interest in the goals of the Federation. They cultivated his friendship, and later a lawyer offered assistance in requesting official recognition. . . . The lawyer drafted the statutes to conform to all legal requirements and deposited them with the governor in Tambacounda. . . .

Unofficially, some observers believed that the lack of recognition had become a political issue. Government officials in Tambacounda and Dakar felt that the Federation was a political organization whose purpose was to thwart the government's development plans in the Bakel region. . . .

Epitaphs (1978–1981) 175

> *The lack of recognition was also compounded by the fact that the Senegalese lawyer who drew up the statutes and deposited them with the government was prominent in the Rassemblement National Démocratique (RND)—until 1981, an illegal opposition party. The Federation claimed no advance knowledge of this relationship until it was too late. In addition, Federation representatives were not entirely clear in conversations with the author about the exact status they sought. Given the previous difficulties between SAED and the Federation, it is not surprising, therefore, that official apprehension about recognizing the Federation existed, that there was fear of political implications, and that the government therefore did not wish to officially respond.**
>
> (David Miller, 'Irrigation and Peasant Autonomy in Bakel', thesis, Cornell University, 1985)

---

connections; anyone who knows our *groupements* is well aware of that. We can guess, however, what pretext has been used.

When we were having great difficulty making our initial application, we were told of a Dakar lawyer who would be willing to help us free of charge. We were glad to hear this, because we had already made several attempts to lodge our application ourselves, and had been dismissed everywhere. We went to see this lawyer and asked him to deposit our application for us; he did so.

Apparently, from what we learned afterwards, this lawyer belongs to an opposition party. That is no concern of ours. We had no contact with him outside that one favour he did us, nor have we had any contact with any politically involved person or political organization whatsoever.

Adrian Adams

During the 1981 dry season, repeated enquiry revealed that the Federation's application for recognition was at the Ministry of the Interior. The official in whose possession it was, told me: 'This is a serious matter, because it involves SAED. You will have to be patient.' Several months later, this same person stated that he had nothing in his files about the Federation; after several further enquiries, he vouchsafed that 'the affair is being handled at the national level', and suggested that the Federation send a delegation to the Ministry of the Interior. Jaabe So wrote to the Minister of the Interior, offering to send a delegation. That letter received no answer.

The French agronomist René Dumont visited Kuŋani in May 1981.

* While this text is inaccurate throughout, endorsing fabrications, emphasis has been added to specific statements wholly and demonstrably false.

Jaabe So was invited to attend a meeting in Dakar between René Dumont and several leaders of Senegalese peasant farmers' organizations. As noted in a report on the meeting for Federation members:

After explaining why he wanted to talk to peasant leaders, René Dumont said that the situation of agriculture in Senegal was very bad. Much of the country's food is imported, food is distributed to people because they can't feed themselves, and the country is not really free because it depends on others to stay alive. True independence means that people can produce themselves what they need in order to live. Up to now, peasant farmers have been regimented (*encadrés*), like soldiers required to obey the orders of a bureaucracy which copies the waste and errors of European society. Aid is wasted, and peasant farmers are exploited. People talk a lot about 'the dams', and say that all will be well in the year 2025. But the dams are not necessarily a solution.

8

A New Beginning?
(1981–1984)

Year Nine: 1981/1982

Adrian Adams

Judging its performance unsatisfactory, funding agencies like USAID and the Caisse Centrale de Coopération Economique, made further aid to Senegal conditional on acceptance of World Bank and IMF demands. From 1979 Senegal embarked on the first phase of a structural adjustment programme requiring reduced government spending, closing down some State corporations and diminishing the scope of others. The first to be closed down was ONCAD, the State groundnut marketing board, which was found to have a deficit of eight thousand million CFA francs. SAED was not closed down, but undertook to entrust more responsibility to peasant farmers, in preparation for its gradual withdrawal.

OMVS began building its dams: work on Diama at the mouth of the river began in 1981.

Jaabe So

N da letara kita, in June 1981 I received a letter from the SAED *Ingénieur-délégué* in Bakel, Pierre Diouf:

> In the future, if you invite us, we should be glad to attend your meetings, in order to give useful information to those present on matters which involve us. Or else you could contact us before your meetings, so that we could provide information on such matters. That would make for better mutual understanding.

Adrian Adams

In October 1981, as a result of meetings with Wyndham James, the new Oxfam field director for coastal West Africa, Oxfam agreed to provide a year's funding for the Federation's proposed literacy programme.

In November 1981 a representative of the Canadian NGO CUSO visited Kuŋani with the French agronomist Claude Reboul and representatives of the Union Générale des Travailleurs Sénégalais en France. As a result of discussions held during that visit, Jaabe So wrote to CUSO about the possibility of funding transport and a starting wage for a mechanic.

Jaabe So

O terinke, o terinke; after waiting and waiting and not hearing anything about our application, we went to Tambacounda. They told us there that the dossier was at the Ministry of the Interior. At the Ministry of the Interior, they said they couldn't find it. So we formed a delegation, with people from Manayeli, Jawara, Yafera, and went to see Gawusu.

He received us in his office, and we told him we'd come about our dossier. He said: 'Your dossier? The man who had it is dead.' I said: 'Did they bury it with him?' I was annoyed. I went to see Tamsir Sall at ENDA [Environnement et Développement Tiers-Monde, an international NGO]. ENDA used to help us when we were in Dakar, they let us use their telephone and typed things for us. Tamsir telephoned the Ministry of the Interior. They said the dossier was there; they hadn't dealt with it, but it was there.

I went back to see Gawusu. By then some Baalu people had also come to see him; they were sitting there. They were talking about a bank account they'd opened, where if you wanted to take out money, you needed SAED's signature. That didn't sound to me like a good idea. I said: 'Gawusu!' He said: 'Yes?' I said: 'I've rung the Ministry of the Interior, they have our papers.'

Then Gawusu told us: 'The reason I haven't helped you so far, is that I think you ought to give up the Federation and form a co-operative instead. If you'd formed a co-operative, you'd already have your papers by now. But *vieux* So is set on a Federation.'

I said: 'What's in it for you? Why are you so keen on our forming a co-operative, that you've refused to help us for seven years? Why are you against the Federation? You don't want an independent farmer's association on the River, that's why. Well, it's what we want. If we get it, fine; if not, that's that. If you help us, fine; if not, that's that.'

I got up and left. After I'd left, he talked to the delegation, telling them they should go in for a co-operative. All that time, the Baalu people were sitting there. Our delegation said: 'We don't want a co-

This study of the Bakel Small Irrigated Perimeters Project was conducted at the request of USAID/Senegal.... The review team visited Senegal between October 30 and November 24, 1981....

When evaluated against the original Project Paper, the project failed and it is apparent that the project emphasis and direction has changed considerably. The reasons and/or basis for these changes are not clear, as they have not been articulated. Furthermore, the project has failed in significantly contributing to the achievement of USAID's general goals for Senegal which include:
1. Increasing food output and raising peasant incomes.
2. Providing a multicrop irrigated agriculture.
3. Allowing the farmer to take over production and marketing factors from SAED.
4. Overcoming the national food deficit.
5. Arresting rural–urban migration by improving conditions in the valley—socially, culturally, and economically.
6. Protecting agricultural production from irregular climatic conditions....

A Region's Social Organization as a Reproducible Development Resource

Short of farming the whole river basin by international agricultural corporations, the hypothesis that existing social organization is a vital resource must be taken seriously....

The original Bakel Irrigation schemes were due to the efforts of well traveled and experienced farmers.... From a social structural perspective, the organizational complexity of Sarakolle civilization and the development orientation of its returned migrants combined with Bakel's physical, political and economic inaccessibility sets the stage for the peasant solidarity movement that takes form in the Federation of Sarakolle villages. If these local organizational resources can be accommodated by national governments, the whole nation may benefit....

Federation input into decision making has been significant... but it is not a routine formal process. Formalization of this process should be done with care not to alienate it from traditional cultural processes. As summarized here and as observed by many scholars and consultants, local social organizations in the Bakel region should be recognized as a valuable and sophisticated resource....

CONCLUSIONS AND RECOMMENDATIONS

... An unfortunate atmosphere has developed regarding the writings of the wife of one of the leading organizers in the region. This controversy is generally over AID and other outside agencies' development policy.

The core of the disagreement is who should determine the development path, the farmers themselves or outside forces, be it government or foreign donor. The basic truth, treated by students of development for at least 30 years, is that all successful development programs must be coordinated with and based on the perceived needs and aspirations of the participants. The participants must be included in planning the development process. To proceed otherwise ultimately ends in failure and occasionally, revolution. One very effective means of achieving this is by listening to the participants. This lesson seems very difficult to learn and it bears constant repetition by those who have access to the ears of change agents.

FARMER FEDERATION/COOPERATIVES

There is a missing link between SAED and the individual perimeter groupments. Rather than SAED mediating conflicts or promoting co-operation between groupments on the one hand, while instilling competition to produce more on the other, the Federations through traditional processes could perform these functions.

Recommendation: Farmer Federations representing existing organizational networks of perimeters cooperatives should be formally acknowledged and incorporated into project management with SAED and USAID....

(Project Review for Bakel Small Irrigated Perimeters, Project No. 685-0208, January 1982)

The Bakel Federation is an example of the very thing that the State and SAED claim to want: that peasant farmers' organizations should master development for themselves. There are two possible solutions, both of which depend upon SAED:

* *either SAED continues to turn a deaf ear and maintains its present attitude, which is, rhetoric apart, opposed to the Federation. At best, it will reinforce peasant farmers' cohesion; at worst, it will crush them. In either case, the unease created in the villages will favour emigration.*
* *or else SAED transforms its attitude towards the Federation. That is obviously the most desirable solution.*

(Confidential note by an OMVS consultant, November 1981)

---

operative. Our president is right to be angry. You ought to help us do what we want.' They all got up to leave. That was when he started telling them: '*Vieux* So is angry with me, but I would really like for us to work together, *collaborer très profondément.*' Then he said Dakar was expensive, and gave them 10,000 CFA. When I heard that, I

A New Beginning? (1981–1984)

wanted to return the money. But the others said: 'Come on, we can use it to buy green tea, it won't stop us doing what we want to do.'

From the very beginning, not only did Gawusu not help us, he did everything he could to hinder us. But we carried on all the same.

Adrian Adams

The dossier was finally found at the Ministry of the Interior, but there were items missing; a new dossier had to be assembled. In May 1982, a new application was deposited at the *préfecture* in Bakel, to be transmitted to the Governor: composed, as before, of two copies of the association's statutes, four copies of the minutes of the constituent general assembly, four copies of the list of founding members, and a handwritten letter of application to the Minister of the Interior, care of the Governor of Sénégal-Oriental.

SAED had encouraged the farming groups to buy oxen and train them, by promising them ploughs and carts; only to announce later that the equipment too would have to be paid for. In order to salvage the scheme, the Federation obtained funding (from ENDA) in order to purchase the equipment on the groups' behalf.

Year Ten: 1982/1983

Jaabe So

Sisoxo Banba-Caalen, Mamadou Cissokho, the development agent from Bamba-Thialène, came to visit the Federation. He first spent two days with Seyidu Ñaŋaane in Baalu; then they came to see me in Kuŋani, saying they'd come to spend the afternoon and night, so that we could talk. I said: *'Bisimilla'*, and killed a sheep for them. When we'd had supper and were drinking tea, Cissokho said he would like me to write down the names of all the people in the farming groups and the area of land they farmed, and send the paper to him. I asked why. He said: 'There's money everywhere.' I said: 'Fine, but we can't do things like that. Money for what? When we have our Federation, and decide what we want, then we'll do what is needed. But we can't become your followers just so you'll find money for us.'

I found out afterwards that he'd said to Seyidu: 'Your president is no good, there's money lying around just waiting to be picked up, why doesn't he go after it? Let's go and see him together; maybe he'll give me some information I can use to help you.' But when they came to see me, they found I had other things in mind; before putting the roof on a house, you need to build the walls.

They left after breakfast the next day. That was when Cissokho put Seyidu in touch with OFADEC [Office pour le Développement Communautaire].

Mammadu Bakari Saaxo

Baalunke lemene ke, that young man from Baalu—what was his name? yes, Seyidu Ñaŋaane—came to see us with a delegation. He said that Jaabe was no use, and SAED was finished; we should go to OFADEC, a new project in Tambacounda. We said we weren't interested. 'If we leave SAED, it will be our decision and Jaabe's. SAED couldn't make us leave Jaabe, let alone you.'

Adrian Adams

Some French students sent by GRDR visited Kuŋani in August 1982 to suggest that it would be a good idea to use grain mills and to sun-dry tomatoes. These were things that had occurred to us before.

In September 1982 Jaabe So went to the *préfecture* to find out what had happened to the application for official recognition. The *préfet* informed him that the Governor had sent the dossier back to Bakel, with a letter saying that it was unacceptable because Article 3 of the statutes, in particular, encroached on SAED's prerogatives as sole agency entitled to conduct development activities in the area. Article 3 states:

> The Federation is a non-political organization, whose essential function is:
> (1) to co-ordinate the various activities of its constituent *groupements*;
> (2) to propagate modern farming methods;
> (3) to help the *groupements* perform a threefold function: production, subsistence and marketing;
> (4) to help the *groupements* to manage their affairs, while respecting their organizational, administrative and financial autonomy.

Al-Haji Denba Bacili

N d'in konten wutu Faransi; in 1979 I left France and returned to Arundu for good. I applied to join the *groupement*, and they accepted me. At that time we were growing maize and rice, and small things like tomatoes and onions.

In 1980 Muusa Bacili fell ill, and God took him to his rest. His vice-president, Mammadu Lamiina Bacili, was made president of the *groupement*. They asked me to become vice-president. At that time I didn't know much about the work; but they insisted.

At that time, there was the Federation, but they were working with SAED. Seyidu Ñaŋaane was the Federation's vice-president. I didn't attend Federation meetings then. One day Mammadu Lamiina called us together, and said that Seyidu Ñaŋaane had told him there was a company in Tambacounda, called OFADEC, that wanted to work with farmers in the Bakel area. They would help us grow bananas.

When we heard that, we were pleased; we're farmers, and if someone wanted to give us banana plants free of charge, that was a good thing. I didn't know then that Seyidu was contacting people on his own, and that the president of the Federation didn't know about it.

When they came back, they wrote a petition to say that if SAED didn't let OFADEC come and work on the River, as of 31 December they would break off relations with SAED. When the paper came to Arundu our president signed it, then the paper went on to Yafera. I thought all the presidents were going to sign it, as far as Gande.

OFADEC gave Seyidu maize, banana plants, and a new HR2 pump. The people in Baalu divided up the maize among themselves. No one in Arundu or the Fula villages knew anything about it, until later on.

Before the OFADEC maize arrived in Baalu, Seyidu sold several tons of it in Bakel; even the people of Baalu didn't know about that. They found out afterwards. That was the first dispute between Seyidu and his *groupement*.

Jaabe So

Piyeeri Juuf ri katt'in ŋa; in November 1982 the SAED *Ingénieur-délégué* in Bakel, Pierre Diouf, came to tell me that the presidents of the Baalu and Arundu farming groups, as well as the small groups SAED had on the *Fan-lenme* near Baalu, had received 50 tons of grain, 760 banana plants, a Lister HR2 pump and 300 metres of pipes from a non-governmental organization called OFADEC, and had written to say that they would break off all relations with SAED at the end of December, unless OFADEC were allowed to come and work in the Bakel area.

I told him: 'They're your relatives; it's your problem. Seyidu betrayed us to join up with you. Now he's betrayed you too. What can we say about that?' Baalu grew only rice, and they'd already harvested it. But Arundu had a field of dry-season maize at the time; they couldn't very well ask SAED for fuel, and the maize dried up and died.

Seyidu took half of the OFADEC grain to Bakel, and sold it there. He sold the other half in Kidira. He kept the money for himself. Arundu received neither grain nor money; nor did Modibo Hamidu, who'd also joined him. As for us in the Federation, we didn't join OFADEC nor break with SAED; we just waited.

Al-Haji Denba Bacili

A da ñi, o da kineyen makka tifi; we'd planted thirty-five acres of dry-season maize. When our president went to SAED for fuel, SAED said they wouldn't give us any unless we withdrew the petition we'd signed. He refused. They said, 'In that case, no fuel.' Our thirty-five acres of maize were ruined, because at that time we didn't have the money to pay cash for fuel.

Then the Federation said that Arundu had betrayed them by joining an initiative that wasn't supported by the Federation as a whole, just by Baalu, Arundu, and the Fula villages on the *Fan-lenme*. It's true, it was our fault; we should have known better. That was how we left the Federation.

Adrian Adams

In October 1982 there was a Federation meeting at the Bakel *mairie*, to discuss how better to organize the work of the farming groups, with priority to subsistence crops, and how better to combine rain-fed and irrigated farming. This was the first Federation meeting I attended. Shortly after the start of the meeting, a man entered the room, exchanging nods with the secretary of the Federation, Ali Tarawore; he did not otherwise greet people or introduce himself, but sat down and began taking notes. The meeting was of course being conducted in Soninke, but most people uneasily began to speak French. When it was finally suggested to the secretary that the newcomer be asked to explain his presence, he said: 'Oh, I thought you knew. *Monsieur* is an Inspector of the *Sûreté Nationale*.'

In early November I was summoned to Kidira, sixty kilometres from Bakel, by the Commissioner of the *Sûreté Nationale*. I was questioned there for four hours: first about my background, my presence in the country, and my everyday life, then about my interest in the Federation's affairs, and my opinion of OMVS plans for River development. I was not told why I was being questioned.

On 16 November 1982, with Jacques Bugnicourt and Tamsir Sall of ENDA, I met with Cheikh Cissokho, now Secrétaire d'Etat aux Eaux et Forêts, in his office in Dakar. At first, he claimed that he knew nothing about my interrogation, nor about the latest rejection by the Governor of Tambacounda of the Federation's application for recognition. But he later appeared to contradict himself, saying that as *responsable politique* for the *département* of Bakel, he knew everything that went on there: I had been heard making subversive remarks at a meeting in Bakel. He added that he saw the Governor every fortnight.

The Government of Senegal, he said, had chosen 'not scientific socialism, but liberal socialism'. The farmers of the Bakel area would have done better to form co-operatives, rather than companies (*sociétés*) which are taxable. There were good villages, like Baalu, which had followed his advice; he had put them in touch with OFADEC.

When asked whether he would support the farmers if they formed *groupements* rather than *sociétés*, he said yes. He emphasized that the Federation was not entitled to seek assistance from abroad; everything must go through the administration.

When Jacques Bugnicourt wrote to him to remind him of his pledge to support the farmers if they formed *groupements*, he replied that he could be of no assistance, as he was not aware of the contents of the dossier.

I also went to see the Director of OFADEC, Mr Carbonare. It was a very brief interview, as he was expecting someone else. He stated that OFADEC intervened only in response to popular demand. The conflict between Jaabe So and Seyidu Ñaŋaane was just a village quarrel; each one wanted to be cock of the walk.

The expected visitor arrived: a blond man in a powder-blue suit, radiating prosperity, later identified to me as the US Chargé d'Affaires. As I left, Mr Carbonare's assistant, Mazide N'Diaye, told me: '*Madame, nous voulons ce que veulent les masses.*'

Jaabe So

N daga komandan ka xadi, in November I went to the Bakel *préfecture* in order to deposit the Federation's application for official recognition. The *préfet* told me he could not accept the dossier, as the Governor had already rejected it.

In December the Governor of Tambacounda, Amadou Thiam, visited the Bakel area, accompanied by a representative of OFADEC. The Governor said at a meeting in Baalu that the area's population should not do anything with OFADEC for the moment, as the government had not yet decided what OFADEC's role should be. He added that he had heard that some peasant farmers wanted to form an independent association. That would never be allowed.

In December, in Dakar, the director of OFADEC told me that the Federation would never be recognized; he knew that for certain, as the Governor of Tambacounda was a personal friend of his. The only way the River farmers could stop SAED taking their land, was by all agreeing to work with OFADEC.

In Dakar, I also met Latyr N'Diaye, head of the Direction de l'Administration Générale et des Collectivités Locales (DAGAT) in the Ministry of the Interior, who advised us to send the Federation's dossier direct to the Ministry of the Interior. We did so, at the end of December 1982.

A *commissaire de police* at the Ministry of the Interior, a friend of Tamsir Sall of ENDA, remarked that we had 'a really thick dossier' in the Ministry's files.

Al-Haji Denba Bacili

N *yaqen watti*, my wife fell ill and I took her to Dakar. One day at the hospital, the person who brought our food told me that a delegation from Baalu was going to call on the Director of OFADEC. It happened that I was staying at the house of a man from Baalu; when I returned there from the hospital, I asked if I could go along to the meeting. I had no business there, I wasn't invited, but as they didn't object, I went along to listen.

The Director's name was Carbonare. When we went to OFADEC, at three in the afternoon, he wasn't there, but his assistant was. The Baalu people spoke. He said in reply that he alone could not bring OFADEC to Bakel. The farmers would have to help him; if we didn't, he wouldn't be able to come.

When I heard that, I said to myself, 'Seyidu Ñaŋaane has got us involved in a bad business.' It seemed that we were to make trouble in order to help OFADEC. That wouldn't do.

Another day, I heard that a Baalu delegation was to call on Gawusu at the *building administratif*. I asked whether I could go along; they said yes. That day Mam Penda, Seyidu's vice-president, headed the delegation, because Seyidu had gone to Baalu with a lorry-load of cement.

While the Baalu people were talking to Gawusu, Oumar Kassimou

Dia rang him up. He was then Director of SAED; this was the first time I heard his name. Gawusu switched the telephone onto a microphone, so we could hear everything they said. He said to Kassimou Dia: '*Justement*, I wanted to contact you, because I've some farmers from Bakel here in my office, and we've been discussing their problem with your people over OFADEC. I asked them to come here in order to make peace between them and SAED.' Kassimou Dia said that would be no problem, because he was going to replace the Bakel *Ingénieur-délégué* with his own nephew Yaya Dia, and give him full powers to do whatever would please the farmers.

Gawusu said he would send a message to the Bakel *préfet*, asking him to call a meeting of the Bakel farmers with SAED, and make peace between them. But Kassimou Dia said there was no need for that; he would tell his nephew to meet with the farmers and come to terms with them.

Gawusu then told the people of Baalu that as foreigners, OFADEC could not go anywhere in Senegal without a safe-conduct from the government. So OFADEC could not go and work with farmers in an area where SAED was already working, without going through SAED. He told them they should not be in conflict with SAED: SAED was our very own, since it was created by the Ministry of Rural Development, whereas OFADEC were foreigners. 'No one can say he doesn't like himself.'

When I returned to Arundu, I asked the president to call the *groupement* together, as I had something to say to them. I told them about the two meetings I had attended in Dakar. I said that the dispute over OFADEC was something we were betrayed into joining. Seyidu Ñaŋaane tricked us into it. The one thing that's our own is the Federation.

Jaabe So

Woten waxati, in February 1983 before the elections, President Abdou Diouf came on a visit to Bakel. I gave him a letter explaining our efforts to win recognition for the Federation. This letter, like all the others we wrote, remained unanswered.

Adrian Adams

An article published in the Dakar newspaper *Le Soleil* of 24 March 1983 stated that the *Fondation canadienne contre la faim* had given

OFADEC 1.26 thousand million CFA francs, to be used for constructing irrigation systems in the *départements* of Matam and Podor.

During a meeting with Jean Collin, special adviser to the President of Senegal, Jacques Bugnicourt of ENDA mentioned the Federation. Jean Collin recommended that a new application be drawn up, with some changes, and deposited at the office of the Governor of Sénégal-Oriental.

A series of articles published in the Dakar newspaper *Le Soleil* stated that all riverside villages in the Bakel area had shown their discontent with SAED by opting to work with OFADEC. At the end of May 1983, *Le Soleil* published a letter from Jaabe So:

> The *groupements* of Yaféra, Kounghani, Bakel, Tiyabu, Manael, Yélingara, Diawara, Moudéri, Galladé and Gandé have not followed Ballou; they have not broken off relations with SAED, nor joined OFADEC. While maintaining working relations with SAED on technical and commercial matters, they pursue as Federation members the task undertaken since 1975: to give priority to food crops (rice, sorghum, maize), and to seek official recognition for the Federation.

The same issue of *Le Soleil* quoted a speech made by a *député* at the National Assembly, suggesting that SAED ought forthwith to be given the means it will require to carry out its allotted task once the OMVS dams have been built.

In May 1983 we wrote to Jean Collin, setting forth in detail the history of the Federation's efforts to gain official recognition, and concluding:

> We are Senegalese citizens; all we ask is that we be acknowledged the right to do what we have in fact being doing for years: taking part in the agricultural development of our area. The spirit in which our Federation was created is fully in harmony with the principles which the government of Senegal claims to uphold. We do not understand why, far from encouraging us, everything is done to discourage us....
>
> We still would like to believe in the rule of law in our country. But in the light of the account above, you will understand, I think, that we sometimes lose hope, and feel that we are being deceived, caught in a web of falsehood and intrigue.

Jaabe So

O giri xadi, in May 1983 a Federation delegation deposited a new application for official recognition (the third since the Federation was created) at the office of the Governor of Sénégal-Oriental in Tambacounda.

A New Beginning? (1981–1984)

Bakel Small Irrigated Perimeters	685–0208
Project Duration	FY 78–FY 84
Total Cost of Project	$6.96 million
Funded to Date	$6.96 million
Project Manager	Joe Salvo
Assistant Project Manager	John Peterson

This project seeks through the development of local institutions to introduce irrigation technology to 24 villages in the Bakel area of the Senegal River Basin. U.S. technical assistance is provided by 3 PSC's working with SAED, the Senegal River Basin's development agency, to carry out the project. During the growing season 1981–1982, the project successfully developed 415 hectares much of which was double-cropped. Rice production averages 5.1 tons per hectare. For the 1982–83 season, the project aims to cultivate 515 hectares of land.

(USAID–Senegal Annual Report)

Ten days later, the *préfet* of Bakel gave me back the application. The Governor had sent it back to him, with a letter to me from Latyr N'Diaye saying that our previous application was being taken care of in Dakar.

Adrian Adams

In June 1983 four *gendarmes*, three of them from Tambacounda, called on me in Kuŋani. The purpose of their visit was to show me a photocopy of a letter sent to the Governor of Tambacounda. The letter, handwritten and misspelt, accused me of 'bringing heavy weaponry (*matériel de guerre*) into the country' and 'meeting Libyan agents between Diaguili and Kounghani'. '*Il faut l'a surveillé.*' It was signed 'Ousmane Diakho'; but the *gendarmes*' investigation showed that although Jaaxo, the equivalent of Tanjigoora, is a very common family name in Kuŋani, there was no Usumaanu Jaaxo in Kuŋani at that time.

When invited to search the premises for *matériel de guerre*, the *gendarmes* demurred. They asked me various questions about my activities, and I made a statement formally denying the allegations contained in the letter. The *gendarmes* then left, saying not to worry. I later lodged a suit for defamation of character against persons unknown.

> Two-thirds of the conseillers ruraux are elected by direct universal suffrage, and one-third by the general assembly of the co-operative or co-operatives operating in the communauté rurale....
>
> (Law 72.25 of April 19, 1972, concerning communautés rurales, Article 4)
>
> Grants of farming land and land for clearing are made by decision of the conseil rural.
>
> (Decree 80.1051 of October 14, 1980, abrogating Article 2 of Decree 72.1288)

Jaabe So

Kineyen waxati, in March 1983 a Federation delegation called on the *préfet* of Bakel to ask about farmers' groups' representation on *conseils ruraux*. The *préfet* wrote to me in June 1983:

> While the regulations provide for representation of *groupements* of an economic and social character, this concerns only those whose existence is legally recognized. And the administration has not yet recognized the Federation.
> Consequently, so long as that situation continues, it cannot take part in the forthcoming deliberative bodies of the *communautés rurales*.
> Conceivably, however, the existing *groupements* under SAED supervision in Lower and Upper Goye could under SAED sponsorship be considered as *groupements* of an economic and social character, and as such lay claim to representation in the *communautés rurales* under the terms of Law 72–25.

Year Eleven: 1983/1984

Jaabe So

O d'a Ministara riyen mugu; in August 1983 we heard that the Minister for Rural Development was coming to Bakel. No peasant farmers' representatives were invited to meet him, nor did he visit any irrigated fields. I heard later on that the only reason he came here, was so he could report Bakel as *sinistré*, a disaster area, and ask for food aid.

Adrian Adams

A four-man commission was sent by the Federation, in a hired car, to visit farming groups which had suffered losses through irrigation problems. Their report was written in Soninke; a French-language version

Gande

We saw their fields. About an acre of maize died because of pump breakdown. Two acres were tilled, but not sown because of this same breakdown.

Siliman Sisoxo (President): 'Our field was ruined through lack of water. The machine broke down about three months ago, and we couldn't irrigate the field. When the machine broke down, we went to tell them in Bakel. Diallo [a SAED mechanic] came. He said there was nothing wrong with the machine, it was just water in the fuel. I said, "In that case, start it up!" He tried to start it, but it wouldn't start. Then he said we needed a new filter. We agreed, and he put it in. He started the motor. The next day, when we went to irrigate the field, the motor wouldn't start. I went to Bakel again. But we were twenty-one days in all without the machine working.

'When the machine broke down again. Nuha [another SAED mechanic] said it couldn't be repaired; they would bring another machine, a new one that had never been used. When they brought the new machine, everyone helped install it. When they tried to start it, it didn't work. They said the pump was no good, but the motor was good. They went to get a new pump.

'After five days, they still hadn't come. I went to find out what was going on; they said they'd gone to fetch a pump from Arundu. They installed the pump from Arundu on the new machine. The next day, while we were irrigating, three bolts came off. We stopped the machine straight away. I went to Bakel the next day. They told me they would come the day after. When they came, they said it couldn't be repaired here; they would need to take it to Bakel. They came back four days later, and took the machine to repair. But the maize had already died. That was the dead maize you saw in the field.'

Yelingara

We saw their fields; in some places, the sorghum crop had dried up. They said the area affected was twenty plots, about an acre and a half.

Waagi Sumaare (President): 'Our field was never levelled; where the sorghum died is higher than the rest, and when we irrigate, the water doesn't reach there.'

Manayeli

We saw their fields. Almost all their crop of sorghum has died. They say that they sowed forty-five acres; the area that has survived is no more than six or seven plots.

Denba Tarawore: 'The problem with our field is that the land was never *levelled. It is very uneven. And the main canal doesn't carry water. When we started irrigating, for five days we couldn't get water to the field. After they [SAED] came, we spent three more days irrigating, and no water reached the field. After eight days in all, we stopped trying. The crop died. We used up two and a half drums of fuel.'*

Bakel-Gassambilage

We saw their fields. They said they don't have enough land; they would like an extension.

Bullayi Ñunma Baxayiro: 'The main canal is no good. It's always collapsing, and we spend a long time without being able to irrigate. Our main problem is water. We take our water from a pond, and we often run out of water.'

Yafera

We saw their fields. The area where the rice crop died because of irrigation problems, is about eight acres.

Laji Timmera (President): 'The canal we built this year won't carry water. That's what ruined the eight acres.'

(Federation report on the 1983 farming season)

---

was sent to the SAED *Ingénieur-délégué* in Bakel and the *Président Directeur-Général* of SAED in Saint-Louis. No acknowledgement was received.

Jaabe So

On 5 April 1984 the *préfet* of Bakel handed me a letter from the Minister of the Interior, dated 14 November 1983:

In response to your letter of 15 September 1983, I advise you that your association, which has not even deposited a dossier making known its existence, must not pursue its activities without a certificate of deposit of such a dossier.

I invited you, by my letter n° 2929/M.INT/DAGAT of 27 April 1982, to deposit a dossier in conformity with current legislation. I remind you that the dossier should be composed as follows:

- A letter of declaration addressed to the Minister of the Interior;
- Two copies of your association's statutes, with a thousand-franc stamp on each typewritten page of one of the copies;
- Four copies of the minutes of the constituent general assembly, indicating the elected officers;
- Four copies of the list of names of founding members, indicating the age, place of residence, nationality and profession of each.

A New Beginning? (1981–1984)

This dossier should be deposited with the *préfet* of Bakel, who will send it to me after an investigation of the morality of the association's elected officers.

I replied on 5 April 1984:

Our association first deposited a dossier in 1981. When we received no answer, I went to Dakar in April 1982, at the head of a delegation. We were then informed that the dossier had been lost after its arrival at the Ministry of the Interior. We were advised to submit a new dossier. We did so. In May 1982, a complete dossier, in conformity with existing legislation as set forth in your letter of 27 April 1982, was deposited at the *préfecture* in Bakel. In August, having called at the *préfecture*, I learned that the dossier had been rejected by the Governor of Tambacounda, because according to him, the Federation encroached on SAED's prerogatives.

We attempted to deposit our dossier for a third time. When I went to the *préfecture* on 23 November 1982, the *préfet* said that he could not transmit the dossier to the Governor of Tambacounda, because the Governor had rejected our previous application. Faced with this impasse, we sent you our dossier by registered post on 29 December 1982, addressed to the Minister of the Interior, with a letter summing up the facts that I have just recounted. Not having received an answer, I wrote to you again on 7 May 1983, to remind you of the matter. In early June, we received through administrative channels a letter from the head of DAGAT, dated 19 May 1983. This letter read in part: 'I acknowledge receipt of your letter of May 7, 1983, in which you express concern about the situation of your association's application for official recognition. In reply, I inform you that the dossier is being processed at our level; we are awaiting the opinion of the Ministry of Rural Development, which is in sole charge of the government's agricultural policy. The dossier was sent there in March 1983. The changes which have taken place in the State have delayed their answer, which is important for the decision to be taken . . .' On 12 December 1983, the present head of DAGAT was so good as to confirm during an interview that our dossier was still at the Ministry of Rural Development, and to promise that we would soon be given an answer. I hope these few facts may be of use to you.

They said in Dakar that the letter from the Minister of the Interior was sent by mistake. But the Federation's dossier that was supposed to be in Dakar, could not be found; we had to make another application. On 20 April 1984 a fourth application for official recognition of the Federation was deposited at the Bakel *préfecture*.

The Minister of the Interior's reply, dated 24 April 1984, was sent through official channels, and reached me in August 1984:

I have taken note of the information contained in your letter of 5 April 1984. However, in order for your application to be properly examined, the best thing would be for you to constitute a new dossier and deposit it with the *préfet* of Bakel, in conformity with the recommendations of my letter n° 7941/MINT/ DAGAT of 14 November 1983.

To enable you to save time, I invite you straight away to remove from your statutes the term 'Soninké' which gives them an ethnic character.

You could use instead a geographical notion such as:
- *Goye supérieur*
- *Goye inférieur*
- *Département* of Bakel
etc. . . .

Al-Haji Denba Bacili

Katta makka ke bice joofe, before the OFADEC maize arrived in Baalu, Seyidu Ñaŋaane sold several tons in Bakel. Even the people of Baalu didn't know about that. They shared out the rest among themselves; it was only afterwards they found out about the maize sold in Bakel. That was the first dispute between Seyidu and his *groupement*.

The trouble between Seyidu and his *groupement* got worse, until they decided to dissolve the *bureau* and elect a new one. He said that as he created the *groupement*, he must remain president. They said he'd tricked them and robbed them too often. They went to the courts, in Bakel and Tambacounda, to start proceedings against him.

The courts decided that whoever won a majority would keep the perimeter. Seyidu received only a small number of votes, from *komo*. So he had to leave the perimeter of which he had been the first president. His vice-president, Mam Penda, was elected president.

When Seyidu was president of the first Baalu *groupement*, all their money was deposited in an account in Tambacounda. When he realized that he was going to be removed from office, he went to Tambacounda and withdrew all the money from the account.

Adrian Adams

After a prolonged internal struggle, the president of the Baalu cooperative, Seyidu Ñaŋaane, was removed from office, having misappropriated co-operative funds including the proceeds of the sale of grain donated by OFADEC. In May 1984, the new president of the co-operative, Mam Penda Ñaŋaane, wrote to Jaabe So giving the results of the vote (he received 356 votes, and Seyidu 264) and asking for Baalu to be readmitted to the Federation. The Federation decided not to readmit them, as it was felt that the Baalu group as a whole was involved in Seyidu's attempts to take control of the Federation and deliver it, first to SAED, then to OFADEC.

A New Beginning? (1981-1984)

Thirty kilometers south of the Federation base of Khounghany... lies the large Soninke village of Balou. Balou, with its irrigation scheme, has become the model site for small-scale irrigation in the Bakel zone and has established a sound reputation basinwide. The village scheme has dynamic, competent leadership, has high levels of productivity, and has maintained its small-scale participatory management style for what has grown into a 108-hectare irrigation scheme.

... For a period of time, the Balou irrigation association president was also vice-president of the Federation because of his contacts outside the Federation with the Ministry of Rural Development, and historically, with the French. Apparently, he did not feel effective as the Federation vice-president and therefore formed a competing organization.

The Federation president was uncomfortable with this internal competition from the dynamic Balou leadership... *The Balou leaders were not only beginning to make their own decisions independently from the Federation, but also to organize other Federation villages around their Balou base.* Appeals were made by the president of the Federation and his wife to AID staff to help sustain the existing Federation leadership in the face of peasant competition.... Following a particularly difficult meeting in which certain Federation members challenged the integrity of the president and thus threatened his position, USAID technicians intervened to help insure that the existing president maintained his position. To help keep Balou in the Federation, a fragile alliance was concluded whereby the Federation president and the Balou president would temporarily serve as coheads of the Federation.

The Balou initiative, however, was maintained and did not stagnate with all the difficulties of the Federation. In 1981, Balou representatives requested formal recognition as a cooperative from the government of Senegal....

The request was granted, and the Balou peasants now work quite independently from SAED. The president of the Federation, meanwhile, has established good working relationships with SAED and USAID....

This chapter has demonstrated that the Federation owes its current manifestation and existence to external support similar to that which it was designed to combat. Key roles have been played by expatriate personnel—both the president's wife and USAID technicians—in supporting this 'indigenous' peasant organization. This support included mediation of internal conflict, permitting the existing leadership structure (her husband) to remain in place. *It also included communication of the Federation's history and purpose in published accounts, communication of the statutes to the Senegalese government, and* direct appeals to AID staff when the existing channels were unproductive. If these efforts had been lacking, essential support would have been withdrawn...

The Federation gave up some of the basic tenants of the participatory model of development in favor of the bureaucratic model implemented by SAED and USAID.... The participatory ideology of development was more closely retained by Balou that eventually rejected the développement administratif and made substantial progress toward the développement paysan so essential to that ideology.*

> David Miller, 'Irrigation and Peasant Autonomy in Bakel', thesis, Cornell University, 1985)

Jaabe So

I d'in tirindi, I was asked at that time to state what I felt about things. I felt that everything was going badly: people were very discouraged, and so was I. The first year things went well, and the second year also. But ever since we'd been under SAED, there had been nothing but problems. When I sat alone at home, I felt very tired.

There was too much trouble: always meetings with SAED, meetings in Bakel, being summoned elsewhere, unanswered letters, lost applications ... As for farming, there were large areas under irrigation, but no help, no advice. The technicians were no good. We did everything by hand, we worked hard, but irrigation wasn't successful, the costs were too high for too little return. That was why many people gave up the work. I kept trying to encourage them by saying: 'It's just because we've not got it right yet.'

The only good thing I could see was that people in the Federation began to realize that peasant farmers should try to run things themselves. The first year, the second year, when we were free to work according to our own ideas, we did well. I reminded people of that, and said that we ought to go back to the beginning, and start working again as we did before; then perhaps things would be better. But we were very tired.

Every year irrigated farming started too late, and *jeeri* farming suffered. That was another reason why people gave up. *Jeeri* farming is very important here. To my way of thinking, irrigation too is important; but the work needs to be done in advance, before it rains. When everything has to be done by hand, repairing canals, tilling the land, it's very difficult. And some things, like building proper irrigation systems and levelling the land, can't be done by peasant farmers working on

---

* While this text is inaccurate throughout, endorsing fabrications, emphasis has been added to specific statements wholly and demonstrably false.

A New Beginning? (1981–1984)

their own. We needed to start all over again. But I was afraid we wouldn't have time.

We heard about OMVS, but no one ever said to the River people: 'We're going to build dams; what do you think about it? What would you like?' No one. The farmers just talked among themselves, and said: 'Why are they building dams without asking us anything?' People would ask me questions about OMVS, and I would say: 'I don't know, I just hear talk.' But we managed to find out things they didn't want the farmers to know; and we were very worried. It surprised me very much that all those researchers and engineers, all those educated people, couldn't see what the result of those dams would be.

When development started here, I made all *chercheurs* welcome. They came to my house, I gave them a room, I gave them food; they did their research. But I saw that the way they went about things, they couldn't tell the difference between what is right and wrong. They just asked for figures: 'How much do you earn? How many people in your house?' They would go into town, meet up with a youth and write down what he told them. One researcher, two, three, four ... When the fifth researcher came, I said: '*Monsieur*, there is no room in my house. Fair enough, you work for a living. But you write too many lies.' We understood that a researcher was an empty envelope; he asked questions, you told him anything you liked and let him go. Researchers like that could not understand our way of farming nor our way of life. They wrote down what they saw with their own eyes; but the country's needs, what could make it go forward, they weren't interested in that.

The *tubab* trained Africans to add up bills for fuel and fertilizers, then they said: 'Here is a technician.' They gave us no new ideas. In an independent country, a technician should teach the peasant farmers. He should discuss things with peasant farmers. He should come to work with the farmers, be there himself to say: 'It's good like this, not like that.' None of the European technicians who came here could say: 'I taught the farmers to do this or that.' And SAED *encadreurs* knew nothing.

Projects never associate peasant farmers with their work. They just collect money and leave. The land remains; it's not developed. They write down the names of peasant farmers, in order to be able to say to financiers: 'We have so many farmers, give us money.'

As a peasant farmer, as a Senegalese, I wanted Senegal to develop in a way that gives Senegal a living; I wanted farmers to be able to stay and work their own land. But that wasn't going to happen with development as I saw it here. We were told at the beginning: 'Peasant farmers don't deserve dialogue.' Ten years later, we were still struggling for our rights, struggling to be recognized.

If Senegal wanted to develop, it should have entrusted the work to the peasant farmers themselves. The government should have respected the *paysans*, and helped them according to their needs: not in their pocketbooks, but in the land. If we had worked like that, in five years our area would have been developed. We didn't need thousands of millions of francs. All those thousands of millions of francs didn't go into the land. Where did it all go? We don't know. What use was it? None.

A development where there's nothing for the farmers, everything for the administration—ah! that development is very negative. When we organized at the beginning, people told me: 'So, if we go in for this sort of thing, the administration will come and make trouble for us.' I said: 'No, when a country develops, the government is pleased; let's just do our work.' I fought hard to get people to organize and work together. Then they came and ruined everything. Technicians were no use to us. Researchers weren't searching for truth. They didn't know that there are living people here.

To develop, you have to work with dignity and honesty, and care about other things than personal gain. I never asked anyone for anything; no one paid me. People in the Federation still wanted to mend things, to try and salvage our ideas from before, even though we'd lost so much time and heart. But in those ten years, we lost a great deal.

PART III

Scissors

9

Mock Reforms
(1984–1986)

Year Twelve: 1984/1985

Adrian Adams

By the mid-1980s the results of the first phase of Senegal's structural adjustment programme, begun in 1979 at the behest of the IMF and the World Bank and supported by the Caisse Centrale de Coopération Economique and USAID, were judged unsatisfactory. A second phase was introduced, to run from 1985 to 1990: the Medium and Long Term Adjustement Plan, which was to cut government spending, freeze wages, raise taxes and prices, and remove subsidies on basic consumer goods.

The New Industrial Policy lifted tax and customs restrictions on entrepreneurs, and made labour legislation less protective of workers' rights. The New Agricultural Policy (Nouvelle Politique Agricole— NPA), introduced in 1984, removed subsidies on seed and fertilizer. Farmers were henceforth to operate on commercial terms, purchasing inputs from the private sector and borrowing money from a newly created National Agricultural Loan Bank (Caisse Nationale de Crédit Agricole—CNCAS). The new measures were initially presented as a new opportunity for peasant farmers. Key words were *désengagement*, the planned phasing-out of remaining State rural development corporations, like SAED; and *responsabilisation*, giving farmers themselves greater control over their own affairs. Legislation introduced in 1984 made it easier for farming groups to obtain official recognition.

In the Senegal River Valley, work began on the OMVS dams: Diama in 1981, and Manantali in 1982. The New Agricultural Policy for the River was presented in terms of *l'après-barrage*, 'after the dams'.

Jaabe So

Ñi mulla, I wanted the Federation to have tractors; so that irrigated land could be ploughed and sown before the rains, and also to show what we could do for ourselves, now that times seemed to be changing. It was not easy to convince people. But finally Monsieur Charles Métras, of the French Mission de Coopération, agreed to help us.

In June 1984, I went to Dakar with two of the people we had chosen to learn to drive our tractors: Mammadu Tuure of Yafera, and Al-Haji Mpali Tanjigoora. At Manutention Africaine, we took delivery of two 46-horsepower Renault tractors, with two disk harrows, two *pulvériseurs* and two trailers. We also met up with a young French technician the GRDR had sent to train our tractor drivers.

The tractors and equipment went by goods train to Kidira, at the Mali border, with the two *tractoristes*. The French technician and I met them there, along with a mechanic from Manutention Africaine, and the tractors were driven to Kuŋani.

With two other people, Denba Goola Jallo of Manayeli and Denba Kulibali of Yafera, and my son Mammadu who returned at that time from his five years' apprenticeship in Dakar, the tractor drivers trained from mid-June to mid-July. My nephew Buna Ja, who drove a tractor for the agricultural co-operative he had helped found at Soboku in Mali, also came to help out for ten days; it encouraged our people that one of them had learnt the work so well.

During that month they ploughed seventy acres, in Kuŋani, Yafera, Mudeeri, Gande, and Jawara. In Kuŋani itself, they were paid to plough fifteen acres of *champ collectif*, my field and Al-Haji Mpali's, and the individual fields of women members of the farming group; they also ploughed some *jeeri* fields, at Lugere, Papata, and Guñan.

We had heard it said that peasant farmers could not take care of tractors properly. To my way of thinking, that shows real contempt. Peasant farmers are human beings, God gave them sense and judgement like everyone else. Our mechanic maintained and repaired the tractors. We trained people to drive them; from the beginning, they kept their worksheets and accounts in their own language. After ten years, they're still working fine, and our tractor drivers are still the best. Everyone here knows that.

Adrian Adams

Having heard about the Federation from Wyndham James, the Oxfam Field Officer for West Africa who funded our experimental literacy programme, Oxfam America sent funds to help equip the Federation's mechanic with tools, spare parts, and transport: a motorcycle for the

TARAKTEERIN GA TUGENE MOXON BE	
What the Tractor Should be Paid (CFA francs)	
To save up the price of two tractors (within 5 years)	2,000,000
To save up the price of the tractor's tools (within 10 years)	550,000
Tractors' upkeep and repairs	1,200,000
Insurance	400,000
Tractor drivers' wages (4 months)	200,000
Expenses	100,000
DUE TO THE TRACTORS' ACCOUNT (yearly)	4,450,000
If we say that each tractor can till 80 hectares a year, 1 hectare should be paid:	
4,450,000 : 160 = 27,812, let us say:	28,000
Let us add fuel and oil:	
30 litres of fuel for 1 hectare:	4,800
and 2 litres of oil:	1,200
TOTAL:	34,000
For 1 hectare, the tractor should be paid	34,000

dry season, and a canoe with an outboard motor for the rainy season, when roads are cut off and only river travel is possible.

At a Federation meeting in May, people had agreed after much discussion that the tractors should charge a price that would cover amortization as well as costs: 34,000 CFA francs per hectare. During their first short season of work, the tractors earned 815,000 francs. Their costs amounted to 255,000 francs, including fuel and lubricant, for spare parts, and 'kola money' for the tractor drivers, assistant tractor drivers and Federation mechanic. The tractor drivers were not paid that first year. It was decided that in future they would be paid 35,000 francs per month worked, and would be entitled to use a tractor to till their personal fields.

When SAED was informed of the Mission de Coopération's plans, it had raised no objection. However, it responded to the arrival of the Federation's tractor by itself offering the services of a tractor for the first time ever—free of charge.

The Bakel *préfet* wrote to GRDR in June 1984, advising them that 'the said "Federation"', not being an official body, was not entitled to contract for the services of a trainer from abroad. The government's

> The point is to create the conditions for a revival of production, in a context favouring the effective participation and heightened responsabilisation *of the rural population at each stage of the development process, consequently restricting the State to a role as catalyst and stimulant.* . . .
>
> (New Agricultural Policy, *Ministry of Rural Development, 1984)*

agricultural policy in the *département* of Bakel could be harmed by such a failure to co-ordinate with SAED. 'I therefore oppose your intervention on behalf of the "Federation", as long as such interventions have not been harmonized with existing official structures.'

GRDR wrote to the Federation enclosing a copy of this letter, adding: 'It would seem advisable for you to meet the administrative authorities in your area and reach a consensus, since as a foreign NGO we cannot intervene against the wishes of the Administration.' Nothing more was heard of the matter.

Jaabe So

O su da ñiiñen gooli, before the 1984 rainy season we all worked together to clear land for an extension to the Kuŋani irrigated perimeter. I went to Dakar; when I came back, I found they had divided up the land among themselves, and had forgotten my share. They offered to divide it up all over again, but I said no, because some people had already started work on their plots. I said I would take the five acres of land at the edge of the perimeter that had been cleared but had no canals. I paid seasonal workers to help me build canals, and paid the tractors to till it.

At the end of August I was summoned to the *préfecture* in Bakel. They gave me back the dossier I had deposited there in April, along with two letters from the Minister of the Interior, dated 24 April and 25 July. The July letter said in part:

In my letter of 24 April 1984, I asked you to remove from the denomination of your association the reference to a 'Soninke zone', which is ethnically discriminatory and furthermore has no basis in any existing administrative entity.

I therefore suggested that you adopt a geographical notion such as '*Goye supérieur*', '*Goye inférieur*', or '*département* of Bakel'; in any case a term without ethnic connotations.

I maintain that position, which is constitutionally justified. . . .

I also inform you that the *groupements* which make up the Federation must themselves be legally recognized before they can federate.

Adrian Adams

In September the dossier was once more deposited at the *préfecture*, with the words '*en zone soninké*' whited out with corrector fluid wherever they appeared.

In November 1984 I was told at the Ministry of the Interior that the farming groups belonging to the Federation should first be registered as *groupements d'intérêt économique* (GIE), as provided for by a recently enacted law. At first the Judge at the Bakel *tribunal* denied that there was any such law. When given the date of its enactment, and asked to look it up in the *Journal Officiel*, he admitted that he was mistaken, but said that the relevant court was in Tambacounda. I travelled to Tambacounda with the vice-president of the Federation. We went to see the *greffier en chef* of the *Tribunal de première instance*. I expected to be told that he was out, that we should not have come to Tambacounda, that we lacked most of the necessary documents, and that the procedure would take at least six months.

As it happened, the *greffier en chef* was in. His name was Cheikh Oumar Diallo. He said: 'I am glad to hear that there is a farmers' association in Bakel. I am from the River myself, and I am in favour of everything that will help its people. As there are eleven *groupements*, and three documents to prepare for each *groupement*, I'm afraid the papers won't be ready by noon; could you come back at three o'clock this afternoon?' In the afternoon, the papers were ready for posting to the Ministry of the Interior.

In September 1984 a team representing a group of French NGOs visited the Federation as part of a study trip to the Valley: among them was Christian Bompard, a Paris businessman, Director of the Centre Lebret, and for the duration of the trip, volunteer chauffeur. They subsequently sent us a copy of their report, which contained a discussion of Federation activities to date: Soninke-language literacy classes, setting up a commission to negotiate with SAED in cases of crop failure, and making available tractors and a qualified mechanic.

Jaabe So

Ken xaaxo ga ñeme, it was at the end of the 1984 rainy season, that I started hearing SAED people talk about the '*périmètre autonome de Bakel*'. I asked what that was, and they gave me a document stating that as of 1 September, SAED had set up a 'SAED–Peasant Farmers Joint Management Committee' (*Comité paritaire*) for the Bakel area, with

TRIBUNAL DE IERE INSTANCE DE TAMBACOUNDA (SENEGAL)
CERTIFICAT D'INSCRIPTION AU REGISTRE DU COMMERCE ET DU CREDIT MOBILIER

The undersigned, Greffier en Chef of the Tribunal de Première Instance de Tambacounda (Sénégal): Certifies that the Groupement agricole de Paysans organisés de Kounghani (Pt. Diabé Sow) *has been entered on the Registre du Commerce et du Crédit Mobilier du Tribunal de Tambacounda, on* Nov. 22, 1984 *under number* 199/A

OBJECT OF TRADE: Dealing in agricultural produce; purchase and sale of various products

This certificate was issued at the Greffe on Nov. 22, 1984 *for the benefit of whom it may concern.*

LE GREFFIER EN CHEF
CHEIKH OUMAR DIALLO

The Federation is beginning to be considered a valid interlocutor by SAED. The fact that the Federation turned down OFADEC's offer to intervene in the zone, and that groupements which had started working with OFADEC were excluded from the Federation, is not unconnected with this improvement in relations, as SAED became aware it had a potential collaborator....

The Bakel Federation has entered an important new stage of its development, by taking concrete steps to master irrigated farming. NGOs which have worked with this peasant farmers' association can testify to its soundness, and the meeting we attended in Yafera showed us that the federative movement is a reality, quite unconnected to any political or self-seeking aims.

In our opinion, the future development of SAED-Federation relations will be an interesting indicator of SAED's will to disengage itself and to promote peasant farmer organizations.

(Rapport de mission au Sénégal, GRDR-IRFED *for a group of French NGOs, August–September 1984*)

---

five farmers' representatives for us and the *Fan-lenme* people and five representatives of SAED, with a SAED-appointed *Directeur de périmètre* in the chair. This was the first I had heard of it. It was discussed at the October meeting of the Federation. On 6 October, I wrote to the new SAED *Ingénieur-délégué* in Bakel, Yaya Dia:

I regret to say that we, the farming groups belonging to the Federation, will not take part in that Committee. We have been in existence since 1974, before

SAED came to the area; as you know, we have continually defended the idea of an independent farmers' organization. To take part in a 'Management Committee' such as you describe might be a step forward for people who had never organized on their own. But for us, it would be a large step backwards. We have progressed much further than that along the road of peasant farmer self-reliance decided on by our government.

If you were to say that you wanted to work with the Federation, for instance by holding two joint meetings a year with us to prepare the two farming seasons, we would be wholly in agreement. For years we have been seeking to work with SAED on terms of equality and mutual respect, so far in vain. You say in your document that you want 'to promote the disengagement of SAED in favour of peasant farmers' organizations'. You should start by acknowledging the existence of those peasant farmers' organizations, instead of pretending they do not exist.

Mammadu Lasana Ba

SAED lost its strength. It was living off USAID funds, but USAID funds couldn't last forever. When that funding stopped, SAED lost its strength.

They said we should change methods, and set up a *comité paritaire*. But when the farmers considered it, they decided they couldn't work with SAED that way, because they already had a Federation. To work in a *comité paritaire* would be as if the Federation had destroyed itself.

Adrian Adams

The head of the French Mission d'Aide et de Coopération visited Kuŋani in November 1984. A member of his party, a Frenchman working for SAED in Saint-Louis, remarked that the Federation *groupements* were the first in the Valley to have registered as GIE.

Mikhael Wane, of SAED in Bakel, was sent to fetch Jaabe So from the fields; in his absence, his companions repeated some remarks he had made in the car. He apparently had said that the Federation was unrepresentative, and that farmers in the area were way behind those of other areas when it came to being ready for *responsabilisation*.

Jaabe So

I senben ga roxo, when SAED realized it was losing ground, they decided to draw up a paper about the Joint Management Committee and have it signed by the peasant farmers. They travelled up and down the river, going to see farming group presidents one by one. They are supposed to be educated people, but I would say from my experience

> *Disengagement from agricultural production implies peasant farmers being made responsible through organizations which take care of their economic and professional interests;*
> - *in order to carry out this process, SAED makes use of all existing forms of association (Co-operatives, village sections, producers' groupements, CUMA (Agricultural Equipment Users' Co-operatives, youth groups...)*
> - *these associations should have considerable economic and decision-making autonomy, and therefore enjoy a legal status affording them, in particular, direct access to credit....*
>
> *For this decentralization to be successful, relations between professional agricultural organizations and SAED should take place within a limited territory. Peasant farmers' decision-making centres should therefore be made closer to those of SAED, by giving the perimeters autonomy....*
>
> *Each of the three autonomous perimeters is provided with a* Comité Paritaire de Gestion *(Joint Management Committee) made up of equal numbers of elected farmers and SAED representatives, responsible for managing shared infrastructure and co-ordinating the services supplied by the perimeter.*
>
> *As this experience calls profoundly into question existing power relationships, care must be taken to see that participants play the game, and training must be initiated to help master management techniques and information (such as functional literacy).*
>
> > (Conseil Interministériel sur les Perspectives et Stratégies de Développement de l'Après-Barrages, *1984*)

of life that they have only a little knowledge, that they use to dominate us.

I asked Mikhael Wane: 'What is this Joint Management Committee?' He said: 'Well, the thing is, when we put something in writing, we send a paper to Saint-Louis, and a paper to the Minister of Rural Development.' I said: 'No; what it really means, is that the *paysans* should sign a paper saying they can't manage things for themselves, and still need SAED. Well, we won't sign it.'

For three months I fought SAED over that. They tried to divide us among ourselves, and to set the Fula against us. They circulated doctored minutes of the meeting held to set up the Joint Management Committee, to make it look as if the Federation was created to discriminate against non-Soninke people. In February 1985 we wrote to Mikhael Wane:

Mr Wane *Ingénieur-Délégué Bakel:*
I thought this whole Federation problem was a thing of the past. Perimeter autonomy is a decision of the government, of SAED, which is a national corporation; it applies throughout its whole zone of activity; if you in the Federation reject it, that means you reject all of SAED's programme as defined by the government; if that were the case, that would be a serious matter, and I must warn you right now that it's impossible. As for the Joint Management Committee, it will be set up as of today....

There can be no question of the Federation dictating our actions... All peasant farmers are equal here, without distinction of race or caste. We cannot accept racial discrimination, we're not interested in Soninké problems and such. Don't forget that SAED belongs to the Senegalese State and must be respected as such.

> (Minutes of the meeting held to set up the Joint Management Committee of the Bakel Autonomous Perimeter, December 1984)

For the past ten years this federation of Soninke peasant farmers has been asking to be officially recognized by the national authorities. Three dossiers have been deposited in succession, with no result as of late 1984. Without official recognition, this fine, truly grassroots-based peasant farmers' organization is at the mercy of regional authorities who can at any time challenge or even destroy these villages' solidarity. At SAED in Bakel, we were told that 'an organization that calls itself "Soninke" cannot be recognized, because national unity cannot tolerate the persistence of divisions based on ethnicity. They must disappear.' Early the next morning, at the guest house where we had been invited to stay, the man who made that fine speech, a Tukulor, brought a cup of coffee to a visiting colleague, a Joola, saying (jokingly, of course): 'Here you are, Joola; you won't be able to say that the Tukulor didn't make you welcome...'.

In fact, the authorities fear the power of this organization, which groups eleven co-operatives like that in Kounghani, for thus organized, peasant farmers no longer accept the authorities' diktats.

> (René Dumont, Pour l'Afrique, j'accuse, 1986. A dedicated copy of the book was sent to Adrian Adams and Jaabe So)

The Federation is an association of farming groups, created before SAED's arrival in Bakel in order to promote agricultural development in our area. Halpulaar and Bambara villages of the nearby *jeeri*—Gabou, Gougnan, Bema, Alahina, Diabal—were originally members of the Federation on the same basis as the Soninke villages by the river. It was SAED which excluded those villages, by stating that it was only interested in irrigated farming....

True, the inhabitants of the riverside villages are Soninke; just as those of a given group of villages in Cayor would most likely be Wolof, and those of a given group of villages in lower Casamance would most likely be Joola. That is a geographical fact, which was reflected in the designation of our Federation as based 'in the Soninke area of Bakel'. When the Minister of the Interior recently advised us that he would like us to remove that designation, we did so without any difficulty. We thought that so-called problem was over and done with. We were profoundly indignant to read in SAED's minutes, in addition to Mr Wane's remarks, the remarks attributed to one of the representatives of the *Fan-lenme groupements*: 'Let me remind the Soninke that all of us here have the same rights and the same duties, we're tired of tolerating the Soninke who monopolize debate at every meeting, to the detriment of other people.' Mr Jeng said nothing of the sort. I can testify to that, as can my colleagues, most of whom understand Pulaar—which in itself is a sign of the close ties we have had since youth with our Pulaar-speaking brothers. As for the other inaccuracies of the minutes, I will mention only a few in passing: the fulsome praise of SAED attributed to Federation representatives; the statement that the *groupements* were sent a note informing them about the joint management committee; the statement that the Federation rejected the joint management committee.

SAED personnel in Bakel know very well that the Federation was set up in order to defend the interests of farmers organized for agricultural development: period. They also know that if the Federation expressed reservations about the creation of this management committee, it was because we would have preferred, in the interests of greater efficiency and openness, that SAED work directly with the Federation as a peasant farmers' organization, rather than through another, rather artificial structure. SAED declined to do so, on the grounds that we were not yet officially recognized. While deploring this attitude of SAED's, which seems to us contrary to the spirit of the new agricultural policy announced by the President of the Republic, we agreed provisionally to provide two delegates—which means that the eleven Federation *groupements* have the same number of representatives as the seven small *Fan-lenme groupements*. We now note that you have written and circulated a tendentious document, the only purpose of which is to discredit the Federation. We can only inform you of our profound indignation at this ploy, and circulate our formal disclaimer as widely as possible.

They went and formed the Joint Management Committee with the Fula. They went to Aruṇdu, the people of Aruṇdu wouldn't receive them. They went to Yafera, the people of Yafera wouldn't receive them. They came to Kuɲani, and I said: 'What's the meaning of this? We aren't pieces of paper. I told you we're having none of it.' So they went on their way. In the end, the Fula dropped out as well.

Mock Reforms (1984–1986) 211

Adrian Adams

In January 1985 I received a letter from a Peace Corps volunteer living in the Bakel area:

> The captain of the military in Bakel wanted to discuss the possibility of doing a fish pond with them. Seems SAED is going to give them 10 hectares to work in the new extension . . . We talked to him and I said I'd do what I could.
>
> I next saw [a USAID technician] who brought me out to the fields to show me an area that he thought would be good for a fish pond. It's an area where a group of women in Bakel have traditionally cultivated rice. It is not in the boundary of the SAED work area. Later I returned alone to the fields and asked a friend about the field. He thought it would be difficult to convince the women . . .
>
> I then went to talk to Wane (SAED). He said to go talk to the *chef de village*. The people I found at the house of the *chef de village* said to go to the *préfecture*. At the *préfecture* I was advised to make a formal request. By the way, Wane told me not to tell anyone that it was the military who wanted to work the field. He told me to tell them it was SAED who wanted to do a demonstration plot.
>
> I returned to Wane and explained what was up. He got all worked up when he heard I went to see the *préfet*. He said to leave them out of it; I would never get the field if I go through them. The next thing he said was what I thought might interest you. He said that *SAED had the right to use any area along the river. The government owns the land and SAED is part of the government so they can take any land they want.* He said to leave it to him and he would get the field. . . .

Jaabe So

O ti, it was decided that now that the farming groups had legal status as *groupements d'intérêt économique*, the Federation ought to take up once more the question of representation on the *conseils ruraux*. On 30 December 1984, we wrote to the Regional Director of the *Service de la Coopération*:

> We recently had the honour of an interview with the National Director of the *Service de la Coopération*, who suggested we get in touch with you. The problem is an important one: participation of our farming groups in the activities of their *communautés rurales* (Mudeeri and Baalu), and in particular their representation on the *conseils ruraux*.
>
> On 7 March 1983, almost two years ago, we wrote to the *préfet* of Bakel, asking for information on the subject. He replied on 22 June 1983, that our farming groups could be represented on the future *conseils ruraux*, with help from SAED.
>
> We wrote again to the *préfet* on 3 December 1984. We thought at that time, on the basis of his previous letter, that our farming groups could be repre-

sented without difficulty, especially as they had in the meantime been registered as *groupements d'intérêt économique*. But the *préfet* replied to our letter on 4 December 1984, saying that in order to be members of the *conseil rural*, *groupement* representatives must be members of a rural co-operative. Never having received any information on this subject, we thought he meant the old co-operatives, and that this meant we were excluded.

On a trip to Dakar, we tried to find out more, in particular with the help of the National Director of the *Service de la Coopération*. If we have understood rightly, as a result of the administrative reform each *communauté rurale* has its rural co-operative, made up of *sections villageoises*; and every farmer belonging to a *groupement* also belongs by right to a *section villageoise*, therefore to a co-operative. Never having been informed of this aspect of the reform, whether at *groupement* or Federation level, we imagined that these new structures had not yet been set up in our *département*. We therefore wrote once more to the *préfet*, saying that it seemed to us our farming groups could form *sections villageoises*. In his reply of 19 December, the *préfet* advised us that the new structures were established in 1983. We were surprised to hear this, as when he replied to our first letter on this subject, in 1983 precisely, he said nothing about any new structures.

Indeed, his recent letter is not clear. Either our farming groups have been included unawares in a *section villageoise*, and will be summoned to the General Assembly of the co-operative of their *communauté rurale*; or—and this is what we fear—our farming groups have been excluded *de facto* from all representation, on the grounds that they did not belong to the old co-operatives. . . .

Our Federation, which has been in existence for the past ten years, represents the majority of organized farmers of the *département* of Bakel, since all *groupements* engaged in irrigated farming belong to it, with the exception of Arundu, Baalu, and the small *groupements* along the Fan-lenme. We had thought that in keeping with the government's new agricultural policy, we ought to co-ordinate our activities as a peasant farmers' organization, with the structures created by the administrative reform. We are unhappy to learn that these structures were set up without us. If we have really been excluded, that means that once more, politicians have denied ordinary farmers their rights as citizens. But we still hope that right will prevail; that is why we have taken the liberty of writing to you and appealing through you to all those who want the new agricultural policy to become a reality.

The Regional Director of the *Service de la Coopération* replied on 8 January 1985:

The *Service de Coopération* is not responsible for your *groupements'* non-representation. It is up to SAED to join the co-operative movement, whose doors are open to peasant farmer organizations recognized by the State. . . .

You wrote too late, after the new structures had been set up and the general assemblies for choosing the co-operatives' councillors had already taken

Mock Reforms (1984–1986)

place.... The Departmental Head of the *Service de la Coopération* tells me that he held a meeting with the population of Kuŋani in front of the mosque, after evening prayers, on 1 October 1983. The population was opposed to the idea of creating a *section villageoise*....

In other words, you cannot, in my opinion, be among the representatives of the co-operative movement in the *conseil rural*, since you have never expressed the intention of joining the co-operative movement.

We replied on 10 March 1985:

You say that the Departmental Director of the *Service de la Coopération* came to Kuŋani in October 1983. We have made enquiries: he did indeed come, and contacted the village chief, who knew nothing about the administrative reform nor about our concern with the matter. He therefore indicated, as he had done before, that he was not interested. In Kuŋani, as elsewhere, the organized farmers were not contacted, whereas they had made known six months previously their desire to join the co-operative structures of their *communautés rurales*.

Your letter makes us realize that in fact, these co-operative structures have nothing to do with farmers.

A copy of this letter was sent to the National Director of Co-operatives in Dakar, as had been a copy of the Federation's initial letter of 30 December. He replied on 22 March:

As you say, the co-operative system is open to every person whose activity is the object of the co-operative, and who resides within the geographical area of the said organization. The same applies to *groupements de producteurs*.

We will do all we can to fulfil your desire to become part of the co-operative structure, in conformity with existing legislation.

Adrian Adams

In January, the Federation was visited by Peter Bloch from the University of Wisconsin's Land Tenure Center (who in 1977, while a visiting professor at the University of Dakar, co-authored the USAID Project Paper on the Bakel Small Perimeters Project). He remarked that from now on, newly irrigable land would be given to progressive individual farmers (*paysans-pilotes*) or private companies. He asked whether people in the area would agree to give up *jeeri* land, and what their reaction would be if the area were designated as a *zone pionnière*.

I subsequently came across a copy of his report. It was consistently inaccurate, stating among other things that it was the Federation which tried to help OFADEC to establish itself in the Bakel area, and that the Federation brought tractors to the area in order to defy SAED policy on introducing animal traction.

> *Grassroots community development must be made possible by economically viable village structures, able to take care of the interests of the producer and the rural collectivity, by means of reform and redynamization of the co-operative movement.... The restructuring operations have been completed. They have led to the creation, at the level of the communautés rurales, of rural co-operatives divided at the grassroots into village units called sections....*
>
> > (New Agricultural Policy, Ministry of Rural Development, 1984)
>
> *In order effectively to support development action of all kinds, which may be undertaken in all economic sectors, it has been deemed opportune to institute the legal framework making it possible to create* groupements d'intérêt économique.
>
> *It is proposed that a simple agreement contracted between two or several physical or moral persons suffice to create and set up for a given period, a* groupement d'intérêt économique *having* personnalité civile, *without it being necessary to form a commercial company, and without any input of capital.*
>
> *This social form will enable the more modest enterprises to organize, and if required, to gain access to credit from appropriate sources....*
>
> > (Preamble to the text of Law No. 84-37 of 11 May, 1984 concerning groupements d'intérêt économique, *published in the* Journal Officiel *of 18 May 1984)*
>
> *The law on* groupements d'intérêt économique, *passed on 11 May 1984, must be considered an integral part of the New Agricultural Policy, even though it was drafted separately. The* groupements *can be freely constituted (provided they make known their rules and regulations and are entered on the* registre du commerce, *if their specific activities give rise to convergent goals and interests. The* groupements d'intérêt économique *are created independently of the co-operative system itself.*
>
> > (Statement on Agricultural Policy *presented by the Government of Senegal to a meeting of funding agencies, June 1986)*

The Federation was also briefly visited in February by a team sent to evaluate USAID's work in Bakel. I saw their report much later, by chance.

I received the George Orwell Memorial Award for 1985, which enabled me to take up a visiting research fellowship at the Institute for Advanced Studies in the Humanities, University of Edinburgh.

A local part-developmental, part-political, part-ethnic organization, called the Soninké Farmers' Federation, was founded under the leadership of Diabé Sow, a migrant returned from France, to act as a countervailing force against SAED. The Federation (which has never received official sanction in a country priding itself on its ethnic harmony) has attempted to exert pressure on SAED by conducting its own public and international relations: having attained the affiliation of nearly all the Soninké villages it encouraged a PVO, which had had success in the Upper Casamance region, to promote alternative crops (bananas) and production techniques, and it obtained a direct grant of two tractors from the French government, in defiance of SAED's policy of promoting animal traction. All this was happening during the first few years of the dépérissement (withering away) of SAED mandated by the contrat-plan (performance contract) imposed by the government for 1982–84. The combination of the contrat-plan's requirements and the Federation's pressure has drastically reduced SAED's presence on the irrigation perimeters in the Bakel region.... Yet the Federation has not succeeded in taking SAED's place. The PVO was rejected when it was realized that their methods were not feasible without enormous subsidies; the tractors have barely been used in the year or more since they arrived; and villages have begun to consider disaffiliation with the Federation.

(Peter Bloch, 'Land Tenure Issues in River Basin Development in Sub-Saharan Africa', Land Tenure Center, University of Wisconsin, June 1985)

Jaabe So

O safandi, in May 1985 we wrote to the Minister of the Interior, telling him the whole history of the Federation's struggle to obtain official recognition. The letter ended:

Monsieur le Ministre, we have taken the liberty of writing to you directly because we are tired of waiting. We represent a farmers' association whose sole concern is the agricultural development of our area. Whoever knows us, knows that. The members of the Federation's *groupements* do not understand at all why their hard work to master irrigated farming is not acknowledged by their government, when other farmers' associations are granted official recognition. In their name and ours, in the name also of the organizations which respect us and want to continue working with us, we most urgently request you, *Monsieur le Ministre*, to give us an answer.

In May, at the suggestion of Sally Ndongo, whom I met then in Dakar for the first time, we also saw Pierre Jacquemot, a special adviser to the

LESSONS LEARNED

1. Projects should carefully assess sociocultural and economic characteristics of a target population during project design. *A project considering introduction of new technology may be more successful if it starts in an area where people are more open and have experience outside of a village or area.*

2. A hands-off approach to a project by a donor can be effective management under certain conditions. *If host country staff is adequately trained, a loose management style can foster a sense of ownership of the project by this staff. This ownership may be an essential ingredient for sustaining key project efforts after donor funding ceases.*

3. Appropriate government policy changes may be necessary to enable target populations to participate effectively in project decision-making. *Although a project may be performing well in the field, if national or regional policies (e.g., pricing) are inappropriate, incentives may be inadequate for project sustainability.*

4. A project management strategy will be more effective if it fosters local participation in management decisions *and permits local organizations to build on indigenous structures and practices in the area. Allowing local organizations to choose their own officers and management style is more likely to foster organizational, hence project, sustainability.*

5. A strategy fostering beneficiary participation requires concomitant and comprehensive training. *Providing responsibility with little means to exercise it may prove frustrating and unproductive. A carefully designed and executed training program can help build beneficiary ownership in project objectives and actions.* . . .

The Federation began as an association to voice Soninke interests regarding irrigated agriculture. Its charter embodied several key concerns regarding its relationship with SAED: . . . Generally, farmers have realized these objectives, partly because of the Federation's efforts. . . .

The Federation, however, has never received formal recognition from the Senegalese government or SAED, although this has been a paramount objective in its dealings with the Ministry of Rural Development. It persists in this effort today. With SAED becoming more amenable to farmers demands, however, farmer support of the Federation has waned. . . .

The new and energetic SAED director in Bakel arrived with a number of concessions . . . His charismatic leadership combined with the major SAED policy changes . . . all contributed to the feeling of a new beginning in the 1983–1984 agricultural season. . . .

Mock Reforms (1984-1986)

As the Manantali dam comes on-stream and the riverine nations... begin to push towards progressively larger perimeters requiring more sophisticated management, one wonders what the status of these perimeters will become.

(Development management in Africa: the case of the Bakel Small Irrigated Perimeters Project in Senegal, *AID Evaluation Special Study, 1985*)

President of Senegal, Abdou Diouf. In June 1985 we wrote to the President:

We have the honour of writing to you about our farmers' association, the *Fédération des Groupements Agricoles de Paysans Organisés du Département de Bakel*.

Enamoured of legality, the Federation's members and leaders have since 1979 made considerable efforts (described in detail in the enclosed dossier) to obtain official recognition: so far in vain.

We now have an idea of where opposition lies: with SAED, with elements of the administration at departmental and regional level—in particular the former Governor of Tambacounda, recently named Director of DAGAT. But we have never seen any valid motive for this opposition.

The Federation has been working with SAED since 1975; its *groupements* were the first to practise irrigated farming in the *département* of Bakel. It has always wanted to take an active part in organizing farming work; this may have created tensions at one time, but should not now, since you yourself have recommended that Senegal's peasant farmers' organizations take on responsibility for their own affairs.

The Federation has no political activity—you must know how our villages vote—and no position based on ethnicity. We have heard rumours that allegations to the contrary have been made against us. Such allegations are wholly false, and we regret that they have never been put to us directly, which would have enabled us to refute them. Whenever possible, we have categorically denied all such insinuations, as the enclosed dossier shows. The Federation is an association of farmers' groups, striving to master the techniques of irrigated farming. That is its only preoccupation.

Your Excellency, we still wish to believe in the rule of law you represent, and in your stated determination to have all active elements among peasant farmers take part in setting right our country's economy. We appeal to you urgently to translate that determination into deeds, and at long last, by recognizing the Federation, show the peasant farmers of Bakel that you include them among your partners.

One day in August, the *préfet* summoned me to Bakel. When I went there, he said: 'Come and sign for your paper.' I said, 'What paper?' He said: 'Your *récépissé*. Come on, sign.' I said: 'You behave as if you're

> The new agricultural policy places a heavy emphasis on intensifying the efforts of the Senegalese nation on the sole behalf of the true actors and beneficiaries of agricultural development, the producers, by freeing them from various negative structural constraints and making them masters of their destinies.
>
> It also clearly expresses the State's great solicitude for the rural world, since it involves creating a propitious setting for the harmonious evolution of peasant farmers' structures: co-operatives and their village sections, as well as groupements de producteurs....
>
> The new agricultural policy will, I am convinced, contribute to restore hope to our worthy rural populations, and lead our friends from abroad better to appreciate our determination to base our development, in the first instance, on our own resources, and therefore to assist us more and better in that task than in the past.
>
> (Introduction by the President of Senegal, Abdou Diouf, to The New Agricultural Policy, Ministry of Rural Development, 1984)

angry with me. Why do you talk to me like that? I'm older than you are.' He said: 'Look, just sign here, keep one copy and give me one.' So I did.

As soon as I had the paper in my hand, I fell ill. I lay there all day, I didn't eat; when people asked me what was wrong, I just said I was ill. I think it came of remembering all the years of struggle to reach this point, and all the disrespect I was shown. They treated me badly. The government treated me badly. In Senegal, democracy is just a word they use. There is no democracy. They humiliate you. They do just as they please.

Thinking about all that, I fell ill. People asked what was wrong, and I couldn't tell them. That was how that paper business ended.

Year Thirteen: 1985/1986

Jaabe So

Koota yogo, one day we were summoned to SAED to be told about the Farm Loan Bank, *Crédit Agricole*. People like the *député* Saada Ja and Manca Njaayi, president of the Mudeeri *Conseil Rural*, spoke first. It's the politicians who make schemes like that work, citizens aren't meant to understand them. Saada said he'd take a tractor and a million francs. Manca said he'd take two *motoculteurs*. All the politicians went along with it.

RECEPISSE DE DECLARATION D'ASSOCIATION

REPUBLIC OF SENEGAL
MINISTRY OF THE INTERIOR
N° 5050 MINT/DAGAT

TITLE OF THE ASSOCIATION
'Fédération des Groupements Agricoles des Paysans Organisés du Département de Bakel'

OBJECT
- To promote solidarity among its constituent groupements;
- to co-ordinate the different development activities of its constituent groupements;
- to help the groupements to improve their mastery of modern agricultural techniques.

OFFICERS

Diabé Saké SOW	President
Ladji TIMERA	Vice-President
Yassa SOUMARE	Treasurer-General
Mamadou DIARRA	Assistant Treasurer
Adrian SOW	Secretary-General

Dakar, 29 July 1985

Le Directeur des Affaires générales et de l'Administration Territoriale
Amadou THIAM

Then they asked me, as president of the Federation, what decision we would take. I said: 'Let me ask you first, what is the Farm Loan Bank's rate of interest?' They said that for every million you borrow, you pay back an extra three hundred thousand. I asked: 'After how long?' They said, at harvest-time.

I told them: 'If I were the government of Senegal, I would forbid that bank. I wouldn't allow it, because it's just putting peasant farmers in gaol. Where is our market? Where can we sell what we grow? What we grow just rots where it lies. We buy from each other, little by little. In any case, we've discussed it in the Federation, and you'll have to excuse us. We won't borrow from the Farm Loan Bank, because if we did, it would just create more problems for us. It wouldn't help us move forward; it would set us back.'

The man said: 'Just what do you think we ought to do instead?' I said: 'You should have farm loans like in France. There you can have a loan for five, ten, even twenty years. But three or four months?' A man who was with them, I think he was from Tambacounda, said: 'That's true. Can a peasant farmer ever make three hundred thousand francs' profit out of a million francs?'

Mikhael Wane accused me of turning the farmers against them. I said: 'We can't stop the Farm Loan Bank from coming to Senegal, but we can decline to join. We are poor people, and we can't afford to go into debt.'

The meeting ended there. Afterwards, everyone here saw what happened to the people who borrowed from the bank. They wanted to get out and couldn't; and no one else joined them. In the end, even Saada Ja said to me, 'Jaabe, what you said was true, but I found out too late.'

Adrian Adams

Construction of the Diama dam was completed in 1986. SAED and CNCAS signed an official agreement. SAED also announced that as of the 1986/7 farming season it would be cancelling the subsidy on fertilizer, previously amounting to over 50 per cent of the price, as well as the interest-free *crédits de campagne*, advances of fuel and fertilizer to be repaid after harvest.

The Cereals Plan increased the price of imported rice, and the price paid to farmers for grain; it announced that fifteen thousand acres would be brought under irrigation in the Bakel area.

In February 1986, the Federation held a fortnight-long training session in Bakel for fifteen future Soninke-language literacy teachers, mostly members of Federation farming groups, with one person from the Sobuku agricultural co-operative in Mali and one from the Projet Karakoro in Mauritania. Training was conducted by two members of a Dakar-based association for the promotion of Soninke language and culture, *Bogu Xura* ('Come out into the open'). Oxfam provided funding for the training session and for starting literacy classes, and for producing and printing a 156-page literacy and post-literacy manual.

In April 1986 the Federation's mechanic held a training session for pump operators and tractor drivers. Six months later, Matforce, the company which has a monopoly of Lister and Hatz motors in Senegal, indicated its willingness to have the Federation represent it in the Bakel area; thanks largely to Christian Bompard, who met the Director of Matforce and offered a personal guarantee. From May to October

Mock Reforms (1984–1986)

1. *Iemenen ma Ieseme*—the child's mother is worried
Reading: *lalle, naani, molle, saase*. Lasana and Laliya. Your child has diarrhoea. Treating illness with water. Stagnant water is dangerous. Heating water in the sun. Rainwater is precious. Maintenance of diesel motors. Numbers: 1 to 1,000. Measures: meters, kilos, litres, money

2. *fulle do xati xaye ri*—the Fula woman brings fresh milk
Reading: *fune, dinde, tafia, yille, xolle, rido*. Yali and Fanta. Feed your child well. Growth, energy, health. Feed your household well. Soil. Fertilizers: rice, sorghum, maize. Ploughing. Planting seedbeds. Drying vegetables in the sun. Dapog rice seedbeds. Protect your harvest against insects

3. *a da begen do cakka do teppuni kita*—she was given cloth, a necklace and shoes
Reading: *bafe, kenne, cakka, peelu, gede*. Account-keeping. Managing a farming group. Clay stoves. Numbers: adding, subtracting, multiplying, dividing

4. *wattun wa jaarene ti haqilen ŋa*—illness is cured by intelligence
Reading: *jexe, wulle, haaja, soxuraqe, ñiiñe, ŋaame*. Fever. Malaria. Drowning. Wounds. Burns. Snakebite. Symptoms of serious illness. Injections can do harm. The family medicine-chest. Minor health problems. Sterility

---

1986, the Federation exchanged a series of letters with War on Want, who sent money at short notice to buy pipes for two standby pumps for emergencies, funded by Oxfam and Oxfam-Belgique (they had been forgotten in the original request), and also sent some Lister tools not available in Senegal—a bearing extractor, an oil seal tool, a flywheel withdrawal DI, and three studs.

Jaabe So

On 1 April 1986 I wrote to the SAED *Ingénieur-délégué*, Firmin Mansis, in Bakel, to ask that in future, SAED work directly with the Federation, rather than working with each farming group separately as before. If that were the case, we could for instance organize meetings on specific dates between SAED and representatives of all the groups belonging to the Federation.

The *Ingénieur-délégué*'s reply, dated 8 April, did not reach me until 28 April:

There is no reason why your wish that SAED work directly with the Federation should not be fulfilled. However, I draw your attention to the fact that your farming groups have received credit from SAED on an individual basis, and most of them still owe us large sums, which they should first repay.

Since I had no answer from the Bakel *Ingénieur-délégué*, on 18 April 1986, we wrote to the Président Directeur-Général of SAED in Saint-Louis:

> Last year, when SAED-Bakel set up its Joint Committee, the Federation told them they would rather SAED worked directly with the Federation. SAED refused, saying that the Federation was not officially recognized.
> The Federation has now been officially recognized, for over eight months. Delegates of our farming groups recently decided they would once more ask SAED to establish direct relations with the Federation; they offered to hold regular meetings with SAED, and named a four-man committee responsible for liaison with SAED. Our letter (copy enclosed) having remained unanswered, they decided that they would not take part in the Joint Committee's forthcoming tour of the area.
> It seems that SAED-Bakel has reacted threateningly, declaring that the Federation wants war, that Federation farming groups will not receive any fuel or fertilizer this year, etc.
> This reaction of SAED-Bakel surprises us greatly. Our association has been in existence for the past ten years; it has been recognized by the administration. We thought that SAED's mission was now to encourage peasant farmer associations to manage their own affairs. SAED-Bakel, on the contrary, refuses to acknowledge the Federation's existence, and wants to force us to accept structures which represent no one; it responds with threats to our suggestion that we work together.
> Threats do not frighten us. We would have preferred, we would still prefer, to continue working with SAED; that is why we are writing to you. If SAED really does not want to work with us, then, God willing, we will continue working all the same, because our work is good for the country.
> We remain at your disposal; if you wish, we will send you a delegation. We hope to receive an answer from you soon.

I had no answer from Saint-Louis. However, the Bakel *Ingénieur-délégué* wrote to me on 27 May:

> I have always admired your frankness and honesty; so I cannot imagine that your letter to the Director had any other purpose than to seek truth and justice. I would therefore be very grateful if you could let me have *a copy of the letter* threatening not to supply the *groupements* belonging to your Federation, or *the name* of the SAED-Bakel employee who uttered that threat.

Enclosed with this letter was a copy of a letter from the Président Directeur-Général of SAED to the Bakel *Ingénieur-délégué*, dated 28 March, in which he stated that the level of debt repayment by farmers

Mock Reforms (1984–1986)

In 1982 and 1985 the BSIP [Bakel Small Irrigated Perimeters Project] was evaluated. The recommendations from these evaluations encouraged USAID to proceed with its plans for an innovative approach to irrigated farming in the Bakel Region. The evaluations revealed that the public sector design and construction of irrigated perimeters in this remote region (sic) and concluded that the potential for private sector support to profitable irrigated farming in the Bakel region was encouraging.... In addition, in 1984, USAID/Senegal welcomed the GOS's New Agricultural Policy (NAP).

It was in this context that the Project Paper for the Irrigation and Water Management Project was first drafted. It took into consideration the recommendations of the BSIP evaluations and the role the private sector was to be given under the New Agricultural Policy. USAID/Senegal authorized the Bakel Project by approving the Project Paper on 20 August 1985, with an original Project Assistance Completion Date (PACD) of 30 September 1990 and a Life of Project (LOP) funding of $8.5 million.

(*Irrigation and Water Management I Project,* Project Assistance Completion Report, *1991*)

Purpose of project: To expand and improve village-level irrigated farming in Bakel, involving greater private sector participation that can be replicated throughout the River Basin....

Senegal has an urgent need for increased food production, for jobs in the rural sector and for increased rural incomes. Currently the country imports about 55% of the food needed in a normal year, has an unemployment rate of over 30%, and has a nearly stagnant GDP although the population growth rate is more than 2.8%. AID is interested in development of irrigated agriculture as one means of addressing these problems....

The New Agricultural Policy of the GOS and the Second Lettre de Mission of SAED have signalled a major change for irrigation development policy; i.e. the state will be getting out of construction and operation of irrigation systems, increasing its effectiveness in development, planning and extension work, and generally reducing public sector personnel. This provides a significantly improved environment in which to develop village owned and managed irrigation systems—the objective of this project....

The Soninke have a strong tradition of community organization and cooperative action which has been a powerful asset in the management of the village irrigation system.... The most prominent feature of all evaluations of the Bakel Small Irrigated Perimeters Project has been the high level of farmer participation and active management role of village leadership. It is this input that has been a key factor in the high produc-

tivity of the village systems and is expected to continue to play an important role as these systems are improved and expanded. . . . Senegal is a functioning democracy and no issues exist with respect to human rights.

AID has been interested in irrigated agriculture . . . since as far back as 1975 when a PWD was prepared for the Bakel Small Irrigated Perimeters Project (BSIP) (initial obligation in 1977 LDP $8.2 million, PACD December 31, 1985) which has constructed irrigation for about 1,200 ha in 28 village systems. . . .

The reconstruction of systems to upgrade performance and facilitate viable operations is expected to include about 400 hectares of the currently established systems. The expansion of systems and construction of new systems will add 800 hectares of irrigation in Bakel Delegation. . . .

The project is technically and economically sound . . . The Mission has considerable experience in Bakel with village level irrigated agriculture. This design draws on that experience, evaluations of the BSIP project, and the numerous studies and reports about irrigated agriculture in Bakel and the Senegal River Basin . . . The IWM [Irrigation and Water Management Project] will be a vital link in accelerating irrigation development along the Senegal River as it improves the village irrigation system and evaluates the scope for alternative technology. . . .

Given the indigenous origins of the groupements in Bakel, and their past record of aggressively dealing with SAED, there is sound reason for being optimistic about the successful involvement of the Joint Management Committee (comité paritaire) in this project.

Farmers participation

From the above highly abbreviated description of the involved institutions, one can sense that the Bakel area is unusual in several ways: (1) the farmers organizations already exist to operate irrigation systems; (2) village level irrigation started in Bakel by individuals acting on their own and not by government action; (3) many groupements were in existence before SAED began operations in Bakel; (4) the groupements are capable of joining forces when their collective interests appear to be threatened; (5) the views of the farmers are presented to SAED on a systematic and regular basis through their groupements, the recently established Joint Management Committee, and in the past by the Federation;* (6) there is considerable cohesion among the members of a groupement and among the groupements themselves; and (7) the two ethnic groups of Bakel area are very open and receptive to new ideas and technology.

---

* Emphasis added. This is the only mention of the Federation in this seventy-five-page-long document.

Mock Reforms (1984–1986)

Conclusion

SAED is improving as an organization... The farmers are well organized... The use of a detailed criteria (people, physical, financial) for the selection of sites for future development... together with more intensive local institution building efforts at the groupement level should make the groupement an effective instrument for development in Bakel. If successful, Bakel's institutions could serve as a model of village-controlled irrigated agriculture.

(Irrigation and Water Management I Project Paper, August 1985)

---

in the Bakel area was unacceptably low, and asked that the farmers be advised of the following measures:

(a) No services or goods will be provided to producers who have not paid off all their debts.
(b) Land will be taken away from peasant farmers who do not pay.

The latest date for debt recovery is set at 30 April 1986.

I wrote again to the *Ingénieur-délégué* on 2 June 1986:

During the Joint Committee's tour of the area, Mr Ndiaga Bèye told the Yafera farmers' group that those who do not work with the Joint Committee will receive nothing from SAED; I was told this by Mammadu Tuure, of Yafera. In Kuŋani, Mr Bèye made similar remarks to me, although not quite so explicit. During a meeting with Bakel-Gassambilaqe, he told them that they would have nothing from SAED; I was told this by Daramaanu Jarra, of Bakel... In Manayeli, Mr Bèye said much the same thing to Manayeli Jallo. I could go on, but I think it is already quite clear. Mr Sy also spoke to Daramaanu Jarra and Buuba Silla in the Bakel market, saying that those who did not go along with SAED would have everything taken from them. Daramaanu asked him if he had not received our letter, and Mr Sy replied, 'We don't need to answer. You've declared war; we'll do the same.' It was after hearing about all this, and not having received any answer to my letter of 1 April, that I wrote to the PD-G of SAED in Saint-Louis, to try and unblock the situation.

On 8 May 1986, the Federation sent Mammadu Bacili of Tiyaabu and Saajo Kante of Yafera as a delegation to SAED. They were received by Mr Sy. I quote from the report Mammadu Bacili sent me right after the meeting: 'We talked to SAED. We discussed the problem of *aménagement* and fuel at length with the man named SY. SAED says that anyone who does not work with the Joint Committee will receive absolutely nothing from SAED. We may as well take a decision now, because things are clear. SAED refuses to work with peasant farmers.' Several times already I have had to calm Federation people down, telling them that SAED has not yet decided on a policy, that we will write to them again.

> In an economy where the survival of the majority of the population depends on the agricultural sector, that sector's contribution to income, employment, foreign currency earnings and public saving must become substantial, for the adjustment programme to be successful. Increasing the national food production will reduce imports, which are already considerable. As for increasing export earnings, it will serve to improve the population's standard of living and increase its purchasing power on the international market.
>
> > ('Presentation of the Agricultural Policy to the Donors' Meeting on Agriculture', Government of Senegal, May 1986)
>
> The middle-term Structural Adjustment Plan is intended gradually to reduce the costs of State intervention, but also implies a clear commitment by the State to take financial responsibility for the public service performed by SAED.
> (a) measures tending gradually to reduce intervention costs are the following:
> • removing all subsidies on fertilizer and irrigation and tillage inputs (from 1984 to 1987);
> • farmers to take on the cost and upkeep of hydro-agricultural schemes and of replacing equipment (from 1987 to 1990 or beyond);
> (b) Support for SAED's public service mission implies the following measures:
> • in the immediate future, payment of the State's arrears to SAED, amounting to 3 billion CFA Francs, in order to clear up its financial position;
> • yearly inclusion in financial legislation, as part of the budget of the Ministry of Rural Development, of the subsidy remunerating the public service performed by SAED (about 1.4 billion a year from 1984/5 to 1986/7)....
>
> > (Conseil Interministériel sur les Perspectives et Stratégies de l'Après-Barrage, 1984)

As for the instructions from Saint-Louis you enclose with your letter: let us leave the land question aside, it is too serious. As for debts—I have always been against going into debt. From the beginning, I advised the farming groups to cultivate a collective field, in order to put money aside to pay their debts. SAED told people that anything collective was no good. It encouraged people to take large amounts of fuel and fertilizer, farm larger and larger areas of land, without giving them technical training, nor trying to find out whether the farmers were making ends meet. I think it would be best if you worked with the Federation, and we all talked with the groups who have debts, to see how they can manage to pay them; rather than breaking off relations, which will harm everyone.

We will come and see you tomorrow, Wednesday, 4 June 1986, to try and work something out.

The next day, I received an invitation to a meeting in Bakel, on 5 June, with the new Président Directeur-Général of SAED. When I was introduced to the PD-G, I told him that I was pleased to see him in Bakel, and that the problems between SAED and the Federation could easily be resolved. He replied that he had not come to talk about the Federation. That was that. Throughout the whole meeting, we weren't allowed to say a single word.

All that time, I'd been telling the Federation people that SAED was going to change, that they couldn't refuse to work with us, because our government is now counting on peasant farmers. After that, I looked as though I didn't know what I was talking about. SAED continued to behave as though we don't exist.

10

What Private Sector?
(1986–1989)

Year Fourteen: 1986/1987

Adrian Adams

The Federation was not invited to the USAID-funded seminar organized in Bakel in August 1986 by the President's Office in order to inform local elected representatives and farmers' organizations about the *après-barrages*.

Jaabe So

I d'o tirindi, the Federation was asked by the Dakar-based Collectif Inter-ONG de Lutte Anti-acridienne (CIONGLA) to organize training sessions and distribution of anti-locust powder in the Bakel area. The situation was very worrying; in August 1986 locusts had already destroyed crops in Jawara and Mudeeri. The Director of Agriculture for Bakel told me he had no fuel. The fuel was in Tambacounda, and Tambacounda refused to send them any; so they'd only placed forty sacks of powder in each of the two *communautés rurales*, which was quite useless.

In September 1986 we organized a training session in Bakel, for twelve riverside farming groups and fourteen inland villages of the *arrondissement* of Jawara. In October the Federation organized delivery of over twenty tons of fenitrothion powder to villages along the river. Between mid-April and the end of June 1987, the Federation distributed seventeen tons of fenitrothion powder in forty-five *jeeri* villages of the *arrondissements* of Jawara, Kidira, and Bala.

CIONGLA wrote to us afterwards that for the 1988 season, they had agreed to only training; the administration would be responsible for distributing powder. Things went wrong from then on.

Everyone knew that SAED was no longer going to supply fuel. The Federation bought ten thousand litres of fuel, held in stock for us at the

About 400 participants were expected at this seminar, representing local elected officials, farmers' groups, sous-préfets and extension agents.

Effective attendance was about 40%; this low rate is due to transport difficulties, heightened by the poor state of the roads at this time of year. The fact that the population is busy with agricultural activities at this stage of the farming season, was also a dissuasive factor.

However, the occasion was distinguished by the presence of eminent personalities, among whom one can mention:
- The Minister of the Protection of Nature
- The Governor of the Region of Tambacounda
- The deputies Sada DIA, Yaya KONATE and Abdoul NDIAYE
- The Mayor of Bakel
- The Consul General of Sweden in Dakar.

(Report on the Bakel seminar on the après-barrages, August 1986)

BP station in Bakel. We agreed that this fuel should be paid for in cash, rather than shared out on credit; that way we could buy more straight away.

Adrian Adams

In September 1986 I wrote to Oxfam America:

It seems that USAID is to fund the post-*barrages* extension of irrigated farming here. The Federation's previous experience of USAID was not a happy one. But things have changed in Senegal since then, and it may be worth our renewing contact with them, if only to try and prevent them from financing schemes destructive of the Federation's efforts.

We called on the then Director of USAID, Mr George Carmer, to suggest that in view of the New Agricultural Policy on *responsabilisation* of farmers' organizations, SAED should work with the Federation. Mr Carmer said that he would put this to SAED, and arrange a joint meeting on his next visit to Bakel.

In January 1987 the Federation received a visit from a representative of the Ford Foundation. He spoke approvingly of the way in which, after the failure of large irrigation schemes in Northern Nigeria, individual farmers had been straggling back to try and set up on their own.

Denba Gise Saaxo

Jaabe t'o n'o ñiiñen raga. Jaabe said we should hold on to our land, otherwise people with more strength than us would come and take it

> The Président Directeur-Général of SAED stressed the importance of responsibilizing management units, promoting and programming the implantation of small and middle entrepreneurs, and relations with agro-businesses.
>
> (Report of a meeting between USAID and SAED, February 1987)

away. We saw that ourselves when SAED gave 375 acres of land in Jawara to Djibi Ndiaye. They first gave him land at Seruka, near Gande; then he decided he didn't want it, and Manca Njaayi agreed to give him the land that we meant for an extension to our perimeter.

Jaabe So

I ga legeri moxon be, in the end whatever SAED decided in Saint-Louis was carried out in the Valley. They started taking people's land and giving it to other people. They were like a new government.

Once SAED and the Mudeeri *conseil rural* gave some land in Jawara to a man called N'Diaye, who worked in Saint-Louis. The Jawara farming group had already applied for that land, but they'd not been granted it. One day they found two bulldozers at work there.

Some people wanted to set fire to the bulldozers. But the president of the farming group, Mammadu Bakari Saaxo, said no. He said they would die for their land if necessary; but they should first go and tell me about it, before doing anything hasty. So they hired a car and came to see me.

As we passed through Bakel, I stopped to see the *préfet*. I asked him: 'Do you know about the land SAED's started work on in Jawara, the land they gave to N'Diaye, who works for SAED in Saint-Louis?' He said: 'No, I don't know anything about it.' We went on to Jawara. I went to see Mammadu Bakari; we went out to the field, and found the bulldozers working there. I held up my hand to them. One of them stopped working, the other wouldn't stop. The first one said to the other: 'Look, SAED is our boss, but these people are organized too. This is their president; let's listen to him. If we don't, it may mean trouble, and I don't want trouble.' I said: 'I'm asking you to stop work until we clear up this business; it's not clear at the moment.'

I went on to Mudeeri; I was told that the president of the *Conseil Rural* was over on the right bank, in Mauritania, at a meeting with Mauritanian officials. I crossed the river. When he saw me, he came over to shake my hand. After I'd greeted everyone, we stood aside to

> SAED is already carrying out a policy of gradual withdrawal in favour of peasant farmers, decentralization, and reducing its operational and investment costs.
>
> It is possible, indeed necessary, to envisage placing redundant personnel in agricultural production, provided that the loss of a paid job be compensated by assistance in the shape of the means to set up a viable agricultural concern, in order to counter the risk of social upheaval. The State might give them plots of land equipped for irrigation. They could be given long-term credit amounting to 4 million francs per hectare, making for (say) 5 hectares each, 8 million francs per person. The State would allow them a lower interest rate. . . .
>
> The withering away of the State corporations in favour of producers organized in powerful co-operatives could take place within a span of 5 years from 1985/6.
>
> An example and goal of progress will be provided by the new type of farmers, former agricultural supervisors turned large or small farmers. These new up-to-date producers and specialized professional farmers, will have a determining effect on the evolution of agriculture towards a modelized and scientifically organized form.
>
> (New Agricultural Policy, *Ministry of Rural Development, 1984*)

talk. I said: 'I've come to see you because there are some bulldozers working on some land the Jawara farming group asked for three years ago. It's right next to their irrigated perimeter, that's not large enough for them. Why did you give that land to someone else?' He said he didn't know anything about it; it must be SAED's doing.

I went to SAED in Bakel. I said: 'They say you sent the bulldozers to work that land. We don't want trouble; but if you don't leave that land alone, things will go wrong.' They sent a message to the bulldozers, telling them to stop work for the time being.

I went back to Jawara, and found the president of the *Conseil Rural* there. I said: 'Look, at eleven-twenty this morning I asked the *préfet* about this business, and he said he'd doesn't know anything about it. At two-thirty I crossed the river to ask you, and you said you don't know anything about it. But we know the land can't have been taken just like that.' He said: 'The trouble with the Soninke is that whenever you grant a piece of land, they start saying it belongs to them.' I said: 'So you do know something about it.'

We held a meeting. I told Mammadu Bakari: 'Don't set fire to the bulldozer, nor assault the driver. Form a delegation and go to Dakar.

Go and see the Minister of Rural Development; if he won't help, go and see the President. If you find they won't behave as human beings, that they just want to do as they please, then come back and die for your land.'

They went to see the Minister of Rural Development, and he rang up N'Diaye in Saint-Louis and told him to give up the land. Afterwards they said they'd incurred expenses working on the land, and Jawara should pay them. There was a meeting. I told Mammadu Bakari it was a lie; no canals were built, and anyway, they hadn't asked for work to be done, they just wanted the land back the way it was. Finally the Minister said to give them the land. They've fenced the land with barbed wire, and they're looking for the means to build an irrigation system there.

Adrian Adams

One of the Federation's literacy instructors began taking notes in Soninke at Federation meetings. These notes were then written out by hand on a stencil and copies printed with a silk-screen for distribution to farming group presidents.

The Federation received another visit from Peter Bloch, an economist from the University of Wisconsin's Land Tenure Center. They were planning a new research project in the Bakel area. I had by chance come across one of his earlier reports, and found it to contain major and potentially damaging inaccuracies; after discussing this, the Federation decided to withhold support from the research project. In February 1987, he wrote to Jaabe So:

We have suggested, and USAID has agreed, that their new project for extending village perimeters in the *département* of Bakel should include a study of the land tenure situation on the irrigated perimeters, and its dynamics since the beginning. For USAID, the point of the study is to help them plan the new interventions on behalf of peasant farmers; for us as researchers, it is to help us understand a highly interesting case of autonomous (*auto-centré*) development; for the River people themselves, it is mainly to be able to recount their lives, and their efforts to achieve survival, to interested persons. I admit that is not of much use to you, but good research does not always have direct and immediate effects. . . . I hope that you will agree that studies such as this do not constitute a threat to the Federation's principles, nor to the villages' independence and self-determination.

Like you, I believe that the experience of the Soninke villagers of the *département* of Bakel deserves to be known, because it has lessons for the rest of the developing world. Like you, I think one should always insist that

FEDERATION MEETING
Jawara, 23 April 1987

SAED:

I da teeni kurunbon dabari Ganden do Gallaade. They have made new fields in Gande and Gallaade. They have started a second canal in Manayeli. But they have done nothing in Jawara. They have not worked on the new field in Kuŋani; the bulldozer destroyed the canals, then left. They have done no work in Yafera.

Bakel has paid 300,000 francs into the bank, in order to obtain two pumps. They have been given one, but have not so far received the other. SAED removed their old float without their knowledge.

SAED has done nothing about the ruined rice-fields, or those which were destroyed by locusts, even though the Federation reported these things a long time ago.

LAND:

Jawara has been given the land it asked for. Yelingara has asked for forty hectares, but the Rural Council has not yet made the grant. Manayeli asked for three hundred hectares between Manayeli and Tiyaabu, but they've not yet agreed on that with the people of Tiyaabu. The Rural Council has given Manayeli one hundred and fifty hectares.

WOMEN'S GROUPS:

Cissoko from the Agriculture Service has been going around holding meetings with women, saying he's going to help them, but he never does. The Federation ought to work more closely with those women, as with the women of Mudeeri. The farming groups should help women more.

ORGANIZATION OF WORK:

People said we should stand up for ourselves, and stop being late starting work every year. We must start work early. The Federation set up a commission, headed by Denba Gise Saaxo (Jawara 1), with Sileymaanu Tuure (Gande), Denba Tarawore (Manayeli) and Bullayi Tarawore (Bakel), to visit all the farming groups and stress the need to start early. It was agreed that the tractors should finish their work, and canals should be repaired, before the end of Ramadan. Where possible, land should be irrigated so that people can begin to till it before the beginning of Ramadan.

Rice seedbeds should be sown at the end of Ramadan; fields should be tilled, and ridges made. Irrigated fields should be sown before it rains.

The president of the Federation said we don't work as hard as we should. Even during Ramadan, if we get up early, we can work before the sun grows hot, to avoid being late this year.

FUEL:

The Federation has been given money to buy fuel tanks and an initial supply of fuel. People said that SAED is not giving fuel to anyone. Those who owe the Federation for fuel should settle straight away so that we can pay for another lot of fuel until such time as our own fuel tanks are installed.

THE FEDERATION'S MECHANIC:

People said that the farming groups are not using the Federation's mechanic as they should; they're still sending for SAED's mechanics. If we don't make use of him, he'll go away; then when SAED goes, the Federation will be in trouble.

The *tubab* know that it is important to have a mechanic; they are helping us to build a workshop. When we start work on the building, each farming group should send two people to help lay the foundations.

GRASSHOPPERS:

People should keep an eye on the ground after the first rains, to see if the eggs left by last year's grasshoppers are hatching. If they see any signs of young grasshoppers, they should tell everyone, so that we can kill them with powder.

---

researchers look for the truth. Like you, I believe that the real experts are the people who live out that truth. Like you, I believe that development is a meaningless word if it is not defined by the people concerned. . . .

Jaabe So

In June 1987 I received a letter from Mr Bloch, announcing his arrival with two researchers. I wrote in reply:

As I said when you last visited here, the Federation has decided no longer to welcome all researchers, but to co-operate only with research that is useful to it. We have found that much of the research done here has not been useful to us, and indeed may have harmed us by circulating false reports.

Adrian Adams

At the end of the 1984 season, the tractors' earnings minus expenses came to 590,000 CFA francs. At the end of the 1985 season, they came to 55,000 francs. At the end of the 1986 season, they came to 2,030,000

> We have not been able to obtain information on all the perimeters. In several cases... it is because the Soninke Federation has been less than eager to accept us as researchers, apparently because of previous bad experiences with being too welcoming to researchers who proved later to have said damaging things about the Federation....
>
> The history of irrigation in Bakel shows that the original dynamic was collective rather than individual. Diabé Sow and others believed that cooperative effort was needed in order to absorb and benefit from the new technology of irrigation.... In the early years SAED put pressure on the GPs [farming groups] to individualize, but there is also substantial evidence that the collective perimeters were running into difficulty independently of that pressure. The principal difficulty is that there are no very good ways of motivating people to work if they feel others are not doing their share....
>
> The new dynamic is for large individual holdings—perimeters established essentially at the behest of and for the benefit of single individuals. The original examples of this type were the perimeters of the two marabouts of Kounghani and Golmi; the most recent examples are Mouderi III and another perimeter in Mouderi which was being developed in 1987 by SAED for a resident of Bakel. These individual farms are a new departure because they will allow the owners to be commercial operators, using modern equipment and agronomic practices and hiring salaried labor....
>
> 'Privatization' is taking place in Moudery and Bakel, and is likely to occur elsewhere, under the aegis of the National Domain Law and subsidized by SAED, i.e. by USAID. It is currently limited to the influential or affluent, who have access to information and to the inner circles of the CR [Conseil Rural]. Does SAED, or does USAID, intend privatization to be so limited, or would it prefer a more democratic approach? If so, then they should work to open up the process of creating individual perimeters to other groups, which may require both education and the provision of incentives to CRs and to local populations.
>
> ('Land Tenure Structure of the Bakel Small Irrigated Perimeters', December 1987)

francs. Two million francs were deposited in a savings account in a Dakar bank. We kept aside 675,000 francs in cash to cover the next season's expenses.

In June 1986 the Federation had asked Gerard Nieuwe Weme of the University of Wageningen, a student researcher with WARDA, to produce a report on the technical problems involved in extending the farming groups' present perimeters. He completed two reports, one on current problems and one on prospects for extension, and presented ten copies of each to the Federation. He found that almost every perimeter had canals that sloped uphill. There were perimeters where certain areas could not be irrigated because they were too high in relation to the canal. This was due either to poor choice of canal location, or to the fact that the land was never levelled. 'In all the perimeters I visited there were problems due to construction faults. That is why from the beginning, farmers have had considerable difficulties with irrigation.'

Jaabe So

I d'in xiri, the Programme Intégré de Podor, directed by Thierno Ba, asked the Ministry of Social Development to invite me to a seminar in Saint-Louis.

Al-Haji Denba Bacili

N gidan towoya, my elder brother fell ill, and it was decided I should take him to the hospital in Saint-Louis. One day near sundown I was sitting by the doorway of the house where I was staying, when I saw an acquaintance from Saŋalu in Mali coming towards me down the street. He said he'd been invited to attend a meeting, but he didn't know where he was meant to go. I said, 'You are welcome to spend the night here; in the morning, we can find out where the meeting is.'

In the morning we went to the Governor's office, and asked a secretary if there was a meeting due to start that day. She said yes, a seminar on *après-barrages* was opening at the Chamber of Commerce at nine o'clock; she explained where it was.

So I attended that meeting. It lasted for three days. It made me understand how important the Federation was. When I went back to Arundu, I told the people there that we must rejoin the Federation.

Jaabe So

In June 1987 I received a letter from the president of the Arundu farming group, asking to rejoin the Federation.

Year Fifteen: 1987/1988

Adrian Adams

Presidential elections took place in Senegal in December 1987, and legislative elections in February 1988. The results were widely con-

A POINT OF VIEW ON THE 'APRES-BARRAGES'

We already know the problems created by irrigated farming. In the *après-barrages*, there will be the same problems, only worse.

(1) *COSTS OF IRRIGATED FARMING*: SAED is to withdraw. Who will ensure the supply of fuel, fertilizer, spare parts? If peasant farmers are not organized, they will be dependent on private traders. As for the basic investments, perimeter layout and pumps, peasant farmers will find it very difficult to take these on unless they are organized. If they are not organized, and do not receive help from NGOs, they will gradually be forced to yield their place to private investors, national or international.

(2) *TECHNICAL SKILLS*: If peasant farmers still do not make enough profit from irrigated farming, this is not just because of the expense; it is also because they do not master the technical skills required: whether in farming, mechanics, or management. Without urgent support for training adapted to peasant farmers' capacities, NGOs' material support will not give good results.

(3) *LAND*: We have been told that peasant farmers will have priority in land distribution within their *communautés rurales*. But we have understood that we need to be vigilant: there can be abuses at the local level. In order to be able to use land to its full value, one has to be able to build irrigation networks. Peasant farmers will certainly not be able to farm all the irrigable land within their village's territory. To whom will that land be given: to peasant farmers from other areas of the country? to private Senegalese entrepreneurs? to foreign companies?

(4) *DAMS*: We know that regulating the river's flow increases the potential for irrigated farming. But the dams were expensive to build: how does the State plan to recover its expenses? Will water for irrigation have to be paid for?

That is how we see the *après-barrages*. The conclusion we draw is that we must *strengthen our organization*. Weak and isolated peasant farmers have no chance of success in the *après-barrages*. Organized in associations, we can share experiences with other farmers all along the river; we can collaborate effectively with NGOs which support development; and we can engage in positive dialogue with the administration.

For a long time, the administration was not favourable to the existence of independent peasant farmers' associations; neither was SAED. We note that things have changed recently. Our presence at this seminar confirms this change, which we hope will continue and gain strength.

(*The Federation's contribution to a seminar on 'The Role of NGOs in the* Après-Barrages', *convened in Saint-Louis by the Ministry of Social Development in June 1987*) tested, and the discontent of opposition supporters sparked off urban rioting. On 1 May, to placate city-dwellers, President Abdou Diouf announced that the price to the consumer of a kilo of rice would be reduced from 160 F to 130 F. Thus while peasant farmers had to pay the full price of fuel and fertilizer on the open market, the price they could hope to receive for their produce was slashed by the State.

Cheikh Cissokho, 'Gawusu', formerly Minister of Protection of the Environment, became Minister of Rural Development in the new government.

The construction of the Manantali dam was completed in 1988. The two dams, Diama and Manantali, together cost nearly two hundred thousand million CFA francs. In 1988, there were over 75,000 acres of peasant-farmer-based irrigated perimeters in the Valley; as against 25,000 in 1982, and 2,500 in 1975. About 72,000 acres of irrigated land were farmed by peasant farmers' *groupements*, and 4,500 acres by small private perimeters, and about 20,000 acres by agro-industries. In 1989 there were 1,165 GIE in the Delta, as against 52 in 1986.

CNCAS replaced SAED as provider of credit to farmers. Borrowers had to deposit 15 per cent of the amount of the loan, and repay the loan with $13\frac{1}{2}$ per cent interest.

Jaabe So

Gilli o ga da gollen joppa, from the beginning there were more women than men in most of our farming groups; they were full of resolve, and worked hard. Often they wanted to grow vegetables in the dry season, but the men in the groups, who had control of funds, wouldn't help them. When we realized that local officials were trying to take advantage of these women, making them false promises and seeking funding in their name, we decided that the Federation ought to help them start up irrigated market gardening in an independent way.

ICCO [Interchurch Organization for Development Cooperation], a Dutch NGO, gave us Nfl. 78,398, 11,633,000 CFA francs, for women's gardens. The first garden was set up in Yafera, in October 1987. The women of Yafera were especially determined, and weren't receiving support from their men. So we fenced two and a half acres of land for them with metal posts and barbed wire; we used our tractors to plough the land and put in canals with our tractors. Then we gave them money for fuel and fertilizer and seed for the first year. We told them that at the end of the season, they should set aside money to cover the next season's expenses.

Adrian Adams

From October to December 1987, numeracy classes were held at farming group level for treasurers, storekeepers, and the future managers of Federation fuel stocks. In January 1988 the Federation held a four-day training session for those who had attended the numeracy classes, on account-keeping, stock-taking and working out a group's seasonal accounts. During the session, various Soninke-language forms were developed, and their use demonstrated using practical examples. During this period, the literacy programme was funded by War on Want.

In February 1988 the literacy programme visited the farming groups, holding meetings to discuss the past season's accounts. There was also a training session for tractor drivers, to evaluate the accounts of the previous two seasons, review management practice, and develop worksheets in Soninke. In April 1988 a meeting was held in Yafera to help the women of the farming group work out accounts for their first season of irrigated market gardening. They found they ought to sell thirty-eight twenty-five kilo sacks of onions, in order to earn the 236,000 CFA francs needed to cover the following season's expenses.

Jaabe So

It had been clear to me for some time that in order to be independent from SAED, the Federation would need to have its own fuel supply and mechanical workshop. Oxfam America helped us to achieve this. We built the workshop during the 1988 dry season, although we didn't equip it until after the rains. We installed two ten-thousand-litre underground fuel tanks, one in Kuŋani and one in Jawara, and signed a five-year contract with BP in Dakar. They filled the tanks in June 1988, just before the onset of the rainy season.

We all met and agreed that we would not give fuel on credit, like SAED. We saw what happened with SAED. People here farm to feed their families, and tend not to sell their produce; so finding cash is no easier at the end of the season than at the beginning. We also agreed that responsibility for pumping fuel from the tanks and responsibility for money from fuel sales should be entrusted to different people. We made special fuel vouchers, entrusted to the person responsible for sales. Fuel was sold only to our own farming groups, in two-hundred-litre drums. In Jawara, the fuel tank was entrusted to Denba Gise Saaxo, and sale of fuel to Yasa Sumaare, the Federation's treasurer. In

> *Of the 823 hectares investigated for the Bakel study, 40% were not irrigated in 1987–89. 15% were technically impossible to irrigate. The reasons given by farmers for not irrigating the other 25% are various and often complicated. Chief among them are lack of money, low yields (partly caused by ineffective irrigation) and mistrust of SAED because of bad experiences in the past.*
>
> > *(G. E. Feenstra, Gestion de l'eau et critères pour la conception des périmètres irrigués villageois de la région de Bakel, WARDA Working Document No. 034B)*
>
> *From 1986 to 1988 the Bakel Project funded the construction of 239 hectares of new irrigation perimeters and the rehabilitation of another 50 hectares of existing systems.... In addition, SAED constructed over 500 hectares of new systems with its own funds between 1986 and 1990. The general quality of the construction and rehabilitation program was substandard.*
>
> > *(Irrigation and Water Management I Project, 'Project Assistance Completion Report')*

Kuŋani, the fuel tank was entrusted to Al-Haji Mpali Tanjigoora, and sale of fuel was entrusted to me.

The Federation sent a delegation to SAED in Bakel, to say that from now on they wanted their own mechanic to work on their pumps. That year, for the first time, we used our tractors to help construct secondary canals, in Gallaade and Arundu.

Adrian Adams

In March 1988 an irrigation engineer from the University of Wageningen, Els Feenstra, began a two-year period of research in Bakel for the West Africa Rice Development Association's (WARDA) *Projet Gestion de l'Eau*. Her comprehensive work-plan, shown to the Federation soon after her arrival, noted that: 'An inventory of the conceptions, ideas and proposals of SAED/USAID and the Federation of Bakel Peasant Farmers, should show what prospects are for cooperating with them'. The final point read: 'When discussions have been completed, and if there is funding, the *aménagements* can be carried out.' Scribbled beneath, in the copy we have, is the comment, 'not by WARDA'.

From 1976 to 1980, SAED constructed 540 acres of irrigated perimeters in the Bakel area; all for *groupements* belonging to the Federation, except for the perimeters for Muslim clerics (in Kuŋani and

> There are a certain number of PIVs [Perimètres Irrigués Villageois] created since 1985 that are ascribed to local politicians or businessmen (commerçants) and their individual families, particularly in the town of Bakel and in Mouderi. ... The first Mouderi perimeter had over 500 members and a good representation of the village population. At that time, collective farming was being practiced. Membership declined rapidly before plots started becoming individualised. Then perimeters were being offered to distinct groups. Mouderi II was for male household heads, with less representation of lower-status individuals; Mouderi III was for the National Assembly depute and his family; Mouderi IV was for the members of the Al-Fala Moslem sect; Mouderi V was for the President of the Rural Council and his allies; Mouderi VI for youth; and Mouderi VII was for women. ...
> Golmi 3 is another case of a politician who has formed a GIE to cultivate his own land. There is also supposedly a private perimeter in Djimbe. Another perimeter (Diawara 2) is based on a sect, the Al-Fallah and is supported financially and spiritually by Arab nations. It is made up almost exclusively of returned migrants and by virtue of the religion, women are not allowed to join.
>> Irrigation and Water Management I Project, 'Midterm Evaluation Report', 1990)

Golmi) and a returned *émigré* in Arundu. During this same period, SAED constructed 188 acres along the *Fan-lenme*.

From 1981 to 1985, SAED constructed 1,250 acres of irrigated perimeters in the Bakel area. Apart from the 345 acres of the new perimeter of Bakel-Kolangal (1982) and the 155 acres of Mudeeri II (1985), these were all extensions to existing perimeters. During this same period, SAED constructed 768 acres along the *Fan-lenme*.

From 1986 to 1988, SAED constructed 1,095 acres of irrigated perimeters in the Bakel area. Of these, 708 acres were for eight new perimeters in Mudeeri, Bakel, and Jawara, only two of which became Federation members. The remainder are extensions to existing perimeters, or new perimeters for existing farming groups, all but 150 acres for Federation members. During this same period, SAED constructed approximately 550 acres along the *Fan-lenme* (figures from G. E. Feenstra, *Gestion de l'eau*).

Jaabe So

O xaalisin ragandaana, the Federation's treasurer Yaasa Sumaare was a member of the Jawara farming group; then he left them to join a new group formed by the Al-Fallah sect. At that time, SAED worked for people like them, and for politicians; they built irrigation systems for them, and gave them pumps. The head of Al-Fallah in Jawara, Al-Haji Seyidu Ba, knew me from the days when he was a trader in Abidjan; he applied for his group to join the Federation. The first Jawara group said they shouldn't be allowed to join, because they were trouble-makers. I persuaded the Federation to admit them, and Yaasa was later reinstated as treasurer. But we soon regretted it.

Njaayi Bullayi, of Mudeeri, had a group of over two hundred women; she applied for them to join the Federation. I asked the Federation to admit them without entrance dues, as they were women. But in the end, we expelled them; they were manipulated by politicians, and didn't respect the Federation's rules.

There was an interpreter at the Bakel courthouse, a *garanke*, who sought help from SAED. SAED told him they'd let him have a machine to work his land. They also arranged with the Rural Council to take someone's land and give it to him. They took some land belonging to Bakel *moodini*; they took so much that they included part of Tiyaabu's irrigated perimeter. They set bulldozers to work there. When the Tiyaabu people saw that, and checked the boundaries of their field, they went to see the *garanke*, and said he should be careful; if he trespassed on their land, that would make more trouble than SAED could deal with. They were working in their field at the time; they said, 'When we finish work, let's go and talk to Mikhael Wane.' The *garanke* had already been to see Wane, and told him the Tiyaabu people wanted their land back. He wasn't pleased with that, so he decided to play a trick on them.

When they came to see him, he said: 'You have hoes and machetes: you've come to attack me! The people of Tiyaabu have come to attack me!' He went to see the Bakel *préfet*. In Senegal, if you have a problem with a government official, they'll always agree against the poor man. The truth doesn't count. So Mikhael went to see the *gendarmes* and the *préfet*, to tell them the people of Tiyaabu came to attack him. They all met, and summoned the people of Tiyaabu. I went to listen. The *préfet* told the people of Tiyaabu they were in the wrong; what they had done was against the law.

That was when I said to the *préfet*: 'Excuse me, as president of the Federation, I have something to say. That land is our perimeter; we chose it before SAED came here, before there was any Rural Council. The farmers had their hoes with them because they'd been working in the fields all day. It's a lie to say they took hoes with them to attack

Mikhael Wane. Mikhael, you know perfectly well they weren't going to attack you. You're just annoyed because they want their land back.

'*Monsieur le préfet*, these are Bacili. That land is theirs. You've done as you like; but God too will do as He likes. The land has spoken; things won't stay like this. Aren't you ashamed as Muslims? Aren't you ashamed before the law? Did the law say to take a field that's being farmed, with the stalks of this year's crop still standing in it, and give it to someone else?'

The *préfet* said he'd summoned them to give them a talking-to, so they wouldn't do the same thing again. I said: 'Talk to Mikhael Wane, and tell him not to take land away from the people who farm it. Tell the *garanke* to take his fence off other people's land.'

Year Sixteen: 1988/1989

Jaabe So

We started up five new women's irrigated vegetable gardens, in Arundu with 140 women, Kuŋani with 70 women, Mudeeri with 230 women, Gallaade with 130 women, and Gande with 70 women. The Yafera garden was in its second year.

We received funds from ICCO for buying fertilizer and building a place to store it in. We talked to a big fertilizer company, SENCHIM, about the possibility of representing them in the Bakel area. SENCHIM said that after a bad experience with a politician in the Bakel area, they decided not to use local agents, but to work through SEDAB, a small company with a representative in Matam. We met SEDAB, and worked out that fertilizer delivered to the farming groups by SEDAB would cost us less than if we bought it at the factory and hired a lorry ourselves. We ordered eighty-one tons of fertilizer.

In May a large Federation delegation went to call on SAED in order to find out what was happening about rebuilding our irrigation systems; that was the one thing SAED was meant to be doing still, and the Americans were supposed to be funding them, but there was no sign of any work being done.

I only knew one of them, named Bill Patterson. He used to come and see me, and say that he wished they could work with the Federation, but SAED wouldn't let them. I said to him: 'Look, you're a black American, I'm a black Senegalese, we're the same; just tell me frankly what you think of SAED. If they were *tubab*, we'd say it's the *tubab* who've ruined things. But look at what we black people do to each

> The project was implemented in two distinct phases: (1) an interim implementation phase from January 1986 through February 1988 . . . and (2) a follow-on phase, from March 1988 to September 1990 with a five-person technical assistance team backed up by short-term consultants, provided by HARZA Engineering Company. . . .
>
> Although the request for proposals for the long-term technical assistance for the Bakel Project was issued on 1 January 86, the contract with the HARZA Engineering Firm was not signed until 5 April 88. This two-year delay was the result of several factors: difficulties in attracting qualified technicians for the very remote and uninviting Bakel Region; scarcity of competent irrigation specialists with French Language capabilities; unacceptable financial proposals; protracted financial negotiations between SAED and HARZA. . . . Amount of contract: $3,285,539. Purpose: Technical Assistance and Feasibility Study.
>
> (Irrigation and Water Management I Project, Project Assistance Completion Report)

other. The Americans gave SAED money on behalf of the poor; and SAED have made sure that the money goes nowhere except their own pockets.' He just smiled. He never worked all the time he was here; in the end, he was glad to leave.

In June, there was a training session for pump operators, held in he new mechanical repair workshop, built with funds from Oxfam America. The workshop was formally inaugurated during that session in the presence of the Bakel *préfet* and other invited officials.

Adrian Adams

A numeracy book was produced, and numeracy classes, including classes for women, took place in farming groups from October to December 1988. In February 1989 there was a training session in account-keeping; reviewing accounts for the grain bank and for construction of a shelter for the Jawara fuel pump revealed mismanagement by those responsible, members of the Al-Fallah group. There was also a training session for women responsible for the new irrigated market gardens.

In March the Federation organized a seminar on ways of improving existing irrigation methods, conducted by G. E. Feenstra, the WARDA engineer working in Bakel, and attended by farming group members responsible for conducting irrigation. Proceedings of the seminar were later distributed among participants. The difficulty was that given the massive flaws in the irrigation systems constructed by SAED, improved management by farmers themselves could have only limited impact.

In addition to G. W. Feenstra, who spent two years in Bakel, WARDA assigned three research students from the University of Wageningen to the area. With the Federation's consent, they each spent six months with a farming group. The student assigned to Gande was not able to produce a plan for perimeter reconstruction, because SAED's topographical work was of poor quality. The other two produced plans for Gallaade and Arundu, fully discussing them with group members. But this was an academic exercise. WARDA's *Projet de Gestion de l'Eau* (in English, Water Management Project) had no access to funds for perimeter reconstruction. The team of consultants from the US-based HARZA group, hired to work in Bakel on USAID's Water Management Project, had no working relationship with WARDA.

Denba Giise Saaxo

The conflict with Mauritania started on 31 March. My elder brother Lasana Saaxo found me in the irrigated fields; I had finished work and had gone down to the river to bathe. He crossed over to the island of Dunde-Xoore, saying that the Fula had let their flocks into his *falo*. Fifteen minutes later I went over to join him. I found that he had chased away all the Fula, except one who had a knife. At that time, the other Fula were on the right bank, ready and waiting: twenty or more people, and they all had sticks.

My brother twisted the Fula's arm and took the knife from him. When the Fula hit him, my brother knocked him down. That was what the others were waiting for.

That day, we were the only Soninke there. They wanted to tie us up, but they didn't. An old woman of about sixty or seventy came up and said in their language: 'Get out! Go away! You've not beaten these two; go away. Otherwise people will come and find you here.' They all left.

My brother took me by the hand and said, 'Let's go.' I was wounded; I had a cut on my head, and hadn't realized it. We came back to Jawara.

A week later, on 9 April, the third day of Ramadan, the Fula let their flocks into Dunde again. When the people from Jawara reached Dunde, they found Mauritanian *gardes* there with the Fula. They chased the sheep out of the fields, and were about to take them back

Originally, it was a returning migrant who provided the impetus for setting up associations, but it very soon became a public matter, with the impetus coming from above. . . .

In the company of Mr W. Patterson, I met the President of the Federation. It is clear that Mr Djabé SOW feels resentment towards SAED, and expresses it undiplomatically. 'SAED has rejected our ideas, a person is not a piece of wood, he has ideas, good or bad . . . SAED operates by force. I was a founder before SAED, for two years; I signed the first contract. There is a lot of American funding, that could have been used profitably. If all the money paid had gone into the land, there would have been a lot of development. SAED has put the farmers in a hole; they are heavily in debt. The irrigation systems were badly built.'

Many reports and studies have been written about the Federation, but I could not find any here. . . . Everything the President of the Federation says is excessive, as if he felt he had been let down. Furthermore, the Federation sometimes does not make it easy for researchers or consultants to work in the groupements. There have been some criticisms of the Federation: it's been said that it's too much of a Sow family affair, because the mechanic is the President's son. It's also been criticized for only federating Soninke, not other farmers. The Federation's tractor charges more than SAED. . . .

The Federation has taken advantage of the opportunity offered by the present situation, with SAED's disengagement: by setting up stores where its members can obtain fuel and fertilizer, it offers a useful service. . . . Up to now, SAED has competed with the Federation by offering its services at subsidized rates.

The Federation seems half-way between a trade union, systematically in opposition, and a co-operative union. . . . It seems to be run relatively democratically, with often lively discussions. The groupements *which belong to it do so of their own free will, and from what their Presidents say, find it to their advantage. The Federation seems open to new members.*

Mr Sow's leadership is undeniable; he has gained the trust of member associations, which respect him, and of funding agencies. . . .

The Federation has been in existence for 15 years; it has survived in a sometimes unfavourable environment. Considering how difficult it is to 'build institutions' and make them last, that is already something. Furthermore, there is at present no other organization offering services to producers, in three of the four zones of the Bakel délégation. After more than 15 years, perhaps the Federation's hour has come.

The President of Moudéry 3, Mr Sada DIA, who is a député, together with the President of Moudéry 7, Mr Manthia NDIAYE, who is President of the Communauté Rurale, sent a circular to all the groupements of the

Bakel délégation on December 26, 1988, suggesting they meet with a view to forming an association. The meetings, dates for which had been set, have not yet taken place because Mr DIA has not been available. Mr DIA has acquired a stock of inputs. It seems that the plan to create an association is a dead letter at the moment: while Mr DIA's plan was mentioned several times in interviews with groupement leaders, no-one, except Mr Manthia NDIAYE, said he would take part in it.

RECOMMENDATIONS

... It would seem necessary to deal with Federation leaders as such, given that it has been recognized by the State.*

Officially, there is a 'Joint Management Committee' for the Bakel délégation. But it has been inactive for several years. It is highly desirable that this Committee should be revived. Generally speaking, what is lacking at present is the co-operative spirit (animation coopérative). Animation coopérative is the responsibility of the Direction de la Coopération, in the Ministry of Rural Development; it is represented in Bakel. Furthermore, SAED already has a policy on co-operatives, as set out, for instance, in the writings of Mr DIALLO, whom I unfortunately sought in vain to meet in Saint-Louis. The suggestions made here should be co-ordinated with the Direction de la Coopération and SAED departments in Saint-Louis.

> (Germain Bertrand, Consultant's Report on Co-operative Activities in the Water Management Project, Phase I, in the Bakel Délégation of SAED, Senegal; Harza Contract n° 58/88. Drawn up for Agricultural Cooperative Development International, Washington, DC; July 1989)

---

across the river. That was when the fighting started. The Fula and *gardes* shot two of our people dead; they took fourteen of them as hostages to Selibabi, and kept them there for four days.

The day the hostages and bodies were brought over to Bakel, the youths said they would take revenge on the Moors living in Senegal. They destroyed the property of Moors and Fula.

It's been five years now. Nothing has been done to define the border. Nothing has been done about our land. We have no standing in Senegal. As far as we can see, Mauritania has won.

There were households from Jawara living in twenty towns in Mauritania. The Mauritanians destroyed everything; the damage was nearly a thousand million. What did the Senegalese government do? It gave

---

* Someone has scribbled on the copy in our possession: 'Careful, the Federation's out of bounds.'

papers to the Fula expelled from Mauritania. Those Fula killed us; now Abdou Diouf has made them Senegalese.

Maybe their grandfathers left Senegal to go and pasture their herds in Mauritania. Their grandfathers and ours, their fathers and ours, lived in peace. Now they've lit a fire between us. The land that we cleared and farmed, they say it's not ours unless we have a paper. That is not Mauritania's land.

I am vice-president of the Jawara 1 farming group, and responsible for the Federation's fuel tank in Jawara.

Jaabe So

Gidimaxa belongs to the Sumaare, the Kamara, the Jabira, the Saaxo; not to the Moors. Senghor gave our land to Ould Daddah; it wasn't Senghor's father's land, it was ours, and he took it from us. We will never forgive him for that, nor the others who took part.

During the conflict, the Mauritanians often fired on Kuŋani; there were soldiers stationed here. I would do the rounds as an *ancien combattant*, to make sure people stayed indoors. One day as I was going past the Friday mosque, I heard the soldier posted on the roof cry out: 'O Mother! *Woyi suma ndey!*' I climbed up there; when I saw him, I came back down to take a woman's cloth that was hung out to dry, and bandaged his leg with it. I said: 'My son, if God has willed your death, you will die; if not, you will live. Don't cry. It's no use crying.' He had a dreadful wound; one of those Iraqi exploding bullets had ground up flesh and bone. They took him to Tambacounda, along those bad roads; he died before they got there.

In May 1989, I wrote to the SAED *Ingénieur-délégué*:

At the beginning of the year, SAED announced a programme of work on village perimeter *aménagements*. Now the rainy season is near, and we note that no work has been done on peasant farmers' perimeters. In the Bakel area, *aménagements* were done only for private individuals: traders, government employees, politicians.

We cannot oblige SAED to work on our perimeters. But we can ask you to tell us clearly if it is in fact the case that SAED no longer does any work for peasant farmers' groups, but only for private individuals.

Adrian Adams

In June 1989, we wrote to War on Want and Christian Aid:

Thanks to funding from Oxfam America, we have been able to install a mechanical repairs workshop and two fuel tanks, which have been in operation for nearly a year; now that SAED has withdrawn from input supply, and with commercial fuel supplies frequently interrupted, these are vital working assets. Thanks to ICCO, we are going to be able to supply the groups with fertilizer. The Federation's literacy and training programme, presently funded by Christian Aid, concentrates especially on practical management problems. Also, an irrigation engineer working for WARDA has gathered a considerable amount of data on irrigation problems of Federation perimeters, which she would like to be of use to us in developing alternative irrigation schemes to those which are being developed without peasant farmers' knowledge and are aimed at removing them from irrigated farming.

For, although the Federation has many assets in hand, the present context is in many ways ominous. In order to defend the future of peasant farmers in River development, that is to say the future of our own people, we will have to make the right decisions: how to plan and seek funding for the *aménagements* we need, how to manage those irrigation systems so as to make them economically viable, how to show a way forward to young people desperately seeking a future. . . .

Jaabe So

In July 1989 I was summoned to a meeting in Bakel with the Président Directeur-Général of SAED, not as president of the Federation, just as an individual farmer, one of eight. I wrote to the *Ingénieur-délégué*:

The only reason for inviting me, would be as president of the Federation. However, you refuse to address me as such. That attitude has lasted for years, and I have no hope of its changing; but I regret it, as being in no one's interest. For years now the Federation ought to have been one of SAED's chief collaborators. Instead, you stubbornly refuse to acknowledge our existence.

Adrian Adams

Loans made by the CNCAS in the Valley in 1988/9 amounted to 550 million CFA francs: 126 millions to farmers in the Middle Valley, the rest to farmers in the Delta.

Yields fell as farmers tried to minimize expenditure on inputs. There were sales of irrigated plots. In some cases, inputs were purchased with advances from better-off people, repaid in kind after harvest at less than the official price; a form of share-cropping which could allow outsiders to invest in irrigated farming without a formal grant of land. In the *département* of Matam, only 121 of the 215 peasant farmers' perimeters (PIV) were partially or totally cultivated in 1989/90.

11

Real Cuts
(1989–1991)

Year Seventeen (1989/1990)

Jaabe So

Ken siine xadi, the Federation helped the women of the Manayeli and Jawara farming groups start their own irrigated vegetable gardens. The gardens in Yafera were in their third year; Arundu, Kuŋani, Gande, and Gallaade in their second. They were doing well. It was only the Mudeeri women's group that had problems. They were too involved in politics.

The SAED *Ingénieur-délégué* had complained to the *préfet* that the Federation wouldn't co-operate within him. The *préfet* called a meeting in his office. In those days, SAED kept the local administration supplied with fuel, so they could shut all doors against other people. But the *préfet*, whose name was Bocar Sy, decided in the end that SAED was in the wrong, because they refused to work with us. He said: 'The government has recognized them; by what right do you reject that decision?'

We built two stores for fertilizer, one in Kuŋani, next to the repairs workshop, and one in Manayeli. ICCO gave us funds for building the stores and purchasing a stock of fertilizer.

Adrian Adams

Literacy and numeracy classes were conducted as usual. It was decided to suspend literacy classes from the following year, as the Federation's limited objective, to assist farming group and Federation leaders to become literate and numerate in Soninke, had been achieved as far as possible; it seemed best to concentrate on putting those skills to good use in managing the Federation's affairs, and on producing reading materials. Materials produced included booklets on measuring lengths and calculating surfaces, and a periodical called *Duna* (The World).

> The relationship between traditional land-use practices, government policies, and current pressures on natural resources is of crucial importance in the struggle for survival of Sahelian agriculture and pastoralism. This project aims to assist those involved in the development of agriculture and natural resources by setting the current situation in its historical and cultural context through the medium of the Sahelians themselves.
>
> (Presentation of the Sahel Oral History Project, SOS Sahel)

The Federation also took part in a project recording the testimony of elderly people about how the environment had changed in their lifetime.

Since fertilizers, unlike fuel, are purchased by individual members, it was decided to place modest stocks with each of the farming groups, before the start of the main farming season. Group storekeepers would sell fertilizer to group members for cash, not on credit. Each year towards the end of the dry season, those responsible for fertilizer at Federation level would visit each group, conduct an inventory of group stores, and collect payment for the quantity of each type of fertilizer sold, taking the group's order for the coming year as well.

Training sessions were held with those responsible for managing the fertilizer distribution system. A first session, in May, introduced a revised set of documents for managing fuel and fertilizer to those responsible for supplies at Federation level; a second session also included farming group storekeepers and treasurers, responsible for ordering and storage at village level. There were also training sessions on fertilizer use.

Jaabe So

We had asked the *préfet* of Bakel some time before, whether it were true, as we had heard, that *groupements d'intérêt économique* now had the same right as co-operatives to representation on *conseils ruraux*; he had not answered us. I wrote to him in about February 1990; he replied that the problem would be carefully considered at the appropriate time.

We had some visitors from a town in France called Bouguenais, that is twinned with the Baalu *communauté rurale*. It was GRDR that suggested they should come to see us. We prepared food and drink for them, and waited on them. They were with the president of the Baalu *conseil rural*, Hamidu Saada Timmera, the son of my father's friend

Saada Sire Timmera, and other local politicians; they ate without speaking to us, then left.

Adrian Adams

One morning in June 1990, a large group of people arrived at Jaabe So's house in Kuŋani. We knew some of them by sight, like the current SAED *Ingénieur-délégué* in Bakel, and the HARZA consultants who had been living in Bakel for some time. Others were new; they introduced themselves as members of a team come from the USA to evaluate the current USAID-funded project in Bakel, and in particular the consultants' work with farmers. They were told that as far as the Federation was aware, the consultants had never worked with farmers at all.

The head of the HARZA party said that they would like to work with the Federation. This was the first time he had ever spoken to us.

Jaabe So

I golliñaŋaana, the gardener who worked for the Americans was a Bacili from Tiyaabu. He told me that the day before they were told to leave, they were saying they ought to work with me. They left behind some seeds for him to give me.

Adrian Adams

The seeds left behind by the Americans were small packets from the USA, several dozen in all: dill, fennel, radish, sunflower . . . Many of them had been marked down from their original price. There was also a small plastic bag full of locally grown marigold seed . . . .

At about this time, we wrote to Christian Aid:

In spite of its achievements, the Federation considers that the present situation is fraught with uncertainty, even danger. The withdrawal of the State proclaimed by the New Agricultural Policy, has not led to support of peasant farmers' organizations; on the contrary, what is being promoted is the taking over of irrigated farming, and the supply of goods and services, by private speculators. It is more urgent than ever that River peasant farmers' associations should meet to discuss a whole range of issues. . . .

These past few years there have been various seminars and conferences on themes such as the aftermath of the dams, to which representatives of River peasant farmers' associations have sometimes been invited. Whatever the goodwill of the organizers, it must be recognized that the style of these gatherings, the presence of large numbers of officials, and the language used,

PROJECT PURPOSE

Expand and Improve PIVs in Bakel

No PIVs have been constructed with project funds since the 1988 arrival of the Harza TA team.... It was found that the inability of the TA team and SAED to accomplish any new construction or rehabilitation was ultimately the result of a stalemate over a private sector strategy, to wit, who should undertake the responsibility.

Development of Replicable PIVs

... The PIVs so far have not been proven to be financially viable and are therefore not replicable in their present form....

PROJECT OBJECTIVES

Participation of the Private Sector

A void exists as far as private sector participation is concerned.... The Federation of Organized Farmers of Bakel is the region's largest private purveyor of services in the areas of plowing, training, and input supply and delivery. Farmer groups who benefit from its services pay membership dues and pay for services on a cash basis. The Federation discourages credit for concern over the risk of debts....

Increase in Commercial Production

Production in Bakel remains one of subsistence....

FINDINGS RELATED TO THE CONCEPT OF THE PIV

The classic notion of the PIV geared towards subsistence production and equitable distribution of benefits is in conflict with the Project's current emphasis on profitability and economic viability....

The Federation PIVs are, on the one hand, the least likely to proceed full force with commercially-oriented production, but, on the other hand, have strong potential to making irrigated agriculture sustainable outside the purview of the Project.

RECOMMENDATIONS BASED ON THE PIV CONCEPT

The selection of PIVs for rehabilitation should not be focused too narrowly on commercial orientation or entrepreneurship but on the identification of traditionally cohesive groups (women, youth groups, emigres)....

In rehabilitating schemes, the design should be tailored to the specific needs and characteristics of the group.... It is recommended that SAED and the technical assistance team consult with WARDA ... on formulating a methodology for participation in design....

In the light of the Federation's pattern of initiatives and program of services, it should be perceived as having a future role in what the Project refers to as the private sector. . . .

Instead of continuing the construction of PIVs on an industrial scale in Bakel, it is recommended that two pilot projects of 50 hectares each utilize the drip/subsurface irrigation method. . . . Drip irrigation technology has been successful in other developing countries like Ivory Coast, Kenya, Mauritius, Venezuela, India to name but a few. . . . Crops include sugarcane, cotton, tomatoes, potatoes, pineapple, strawberries, artichokes, asparagus, bananas, and even fruit and nut trees like mangoes, oranges and macadamia nuts, to name but a few.

> (Irrigation and Water Management I Project, Midterm Evaluation Report, July 1990)

Following five years of implementation, the project failed to achieve its major objectives: there was no private sector involvement in irrigation development and agricultural services, a financially viable irrigation prototype was not developed and replicated by farmers, and no technical improvements were made in the quality of the design and construction of the project-supported irrigation systems. As a result of this unfortunate situation, farmers' incomes, employment and agricultural production did not increase in the Bakel area.

Based on the recommendations of the June 1990 mid-term evaluation, the Mission and the Government of Senegal (GOS) determined that it was no longer desirable to continue implementing the project. On 4 October 1990, the Mission and the GOS formalized their agreement of early termination of the project which took effect on 31 March 1991. . . .

A total of 142.3 man-months of long-term technical assistance were provided under the SAED/HARZA Contract.

The HARZA contract provided 20 person-months of short term consultancies and 6.5 person-years of long-term staff services under four subcontracts with U.S. firms for a total cost of $1,366,546. These were: (1) Agricultural Cooperative Development Int. (ACDI) (2) Associates in Rural Development (ARD) (3) Blue Nile (4) DEVRES. . . .

The first technicians that were mobilized were the Chief of Party/Design Engineer and the local-hire Senegalese Irrigation Operations Engineer. HARZA dismissed the former after four months and the latter after six and a half months. The replacement Chief of Party arrived in the fifth month. The replacement Operations Engineer, a recent university graduate, with very limited French language skills, did not arrive until the thirteenth month of the HARZA contract. During their respective assignments neither one was active in field operations. Although the Chief of

Party, a PhD in agricultural engineering, operated as an irrigation design engineer, his field and management experience was very limited. Furthermore, he was frequently involved in overall team management and administration until the end of the contract, which prevented him from effectively accomplishing his technical goals. The Operations Engineer was oriented towards developing computer models and spent most of his time in the office. As a result of all this, field operations, systems maintenance and technology transfer were never given priority under the host-country contract. Furthermore, the technical assistance team was unable to prepare in a timely fashion design and construction standards acceptable to either the Mission or SAED. The contract Agronomist was successful in revitalizing the SAED demonstration farm; the farm's extension program, however, was started late and had a limited impact on irrigated farm operations. The Rural Development Specialist and the Administrator were less successful in technology transfer as they hardly provided any significant training either to SAED staff or project farmers....

The HARZA Contract was completed as scheduled, on 30 September 1990. As of this report, HARZA's total claims were $3,406,579 against total contract costs of $3,842,728.

(*Irrigation and Water Management I Project,* Project Assistance Completion Report, *1991*)

---

have meant that peasant farmers' representatives have played only token roles. True, at least they have met, and said that they would like to meet more often. But once returned home, they are once more submerged by their daily cares. To be able to meet and exchange views, requires organization, time and money.

Year Eighteen: 1990/1991

Jaabe So

Ke me-ñiye su, with so many meetings and training sessions, we needed a place where up to twenty or so people could meet and spend the night. With help from War on Want we built two large rooms opening onto a veranda, next to the workshop and fertilizer store.

There was quite a lot of money left over from what Oxfam America gave us for installing the workshop and fuel supply. They authorized us to use it for buildings to house the rice huller and grain bank and a ventilated storeroom for onions, and a wall all around, to make a proper compound.

Jaabe So

The SAED *Ingénieur-délégué* in Bakel sent me a letter, addressed for the first time ever to 'Monsieur le Président de la Fédération des Soninkés':

I take pleasure in informing you that two peasant farmers who are members of your organization have been chosen to make a trip to Niger.

They have distinguished themselves by their assiduity at work, the high yields they have achieved in irrigated farming, but especially their altruism on behalf of their *groupement*.

Monsieur Yassa Soumaré: Diawara II, Madame N'diaye Boulaye N'Diaye: Moudéri VI.

It was decided that Federation people should not take part in the trip to Niger, because of SAED's refusal to work with the Federation. Yaasa Sumaare went anyway; SAED was giving each person who went 60,000 CFA francs. They were delayed on the way back, and he wasn't there at the next Federation meeting; so we found out.

Adrian Adams

At the Federation meeting in January, all those present had been given individual nomination papers, to be sent in sealed envelopes by a given date to the delegate of the Bakel farming group. When the envelopes were opened it emerged that the only new candidate was Al-Haji Denba Bacili, whom many people had nominated as vice-president.

However, at the following meeting in February, it was learned that the treasurer, Yaasa Sumaare, who was not present, had gone on the trip to Niger organized by SAED. It was decided there and then to remove Yaasa Sumaare from his position as treasurer, replacing him with the then *Commissaire aux Comptes* and appointing a new one. For the other positions, it was decided that as a majority of the nomination papers favoured the new candidate for vice-president, there was no need for a further election.

Jaabe So

We found out that Yaasa Sumaare had been advancing fuel to the Al-Fallah group, both for irrigation and for their mosque's generator, using the Federation's *caisse* to make up the total when it was time to order a new supply. He said he did it to help the poor.

For all the talk of *désengagement*, and the number of people it had let go, SAED was still around. In June that year there was an article in

Real Cuts (1989–1991) 257

> In the département of Bakel, the first visit was for the federation of organized farmers in Kounghani, an inter-village project headed by Mr Diabé SOW, an emblematic figure among migrants. There was a moving speech, recalling the federation's history and its constant struggle to remain in control.
>
> In the town of Bakel, a throng welcomed the visitors to the women's communal vegetable garden. They were joined by Mr Cheikh Abdoul Khadre CISSOKO, Minister of Rural Development and Hydraulics.
>
> This is a magnificent garden of several hectares, its infrastructure funded by the Caisse centrale de coopération économique.
>
> (Actualité Migrations, published by Office des Migrations Internationales [OMI, a French government agency], March 1991)

the newspaper *Walfajri* all about how peasant farmers in the Valley couldn't manage without SAED, how the withdrawal of the State had 'left room for a pack of exploiters'. We wrote them a letter, which they didn't publish:

> You say that peasant farmers left alone and resourceless by the withdrawal of the State are having a hard time and sometimes fall prey to crooks. But your journalists see no solution other than that suggested by the SAED workers' union, namely reviving SAED, 'whose experience makes it uniquely capable of guiding the *après-barrages*'.
>
> If, as you say, the peasant farmers 'are badly trained' and 'cannot yet manage things for themselves', whose fault is that, if not that very same SAED which supposedly guided them for over twenty years?

Adrian Adams

Claude Evin, then French Minister for Social Affairs and Solidarity, visited Kuŋani in February 1991, accompanied by the French Ambassador to Senegal, the Head of the French Mission de Coopération, the Director of the Caisse Centrale de Coopération Economique, the representative of OMI in Senegal, and the *chef de projet* of GRDR in Senegal. I read out a speech we had agreed on beforehand, to which they made no reply. It ended:

> We would like to say to you clearly, that you who are concerned with migrant workers' reintegration should talk to us, the Federation's leaders. You may say it takes time to speak to people with whom you have no common language. But so much time has been lost already! Take time not to lose more.

Jaabe So

If the money they gave Gawusu had really been put into the Bakel women's market garden, it would have been enough to keep it going for twenty years. But all that was done was to bury some pipes and install an old pump. When the visitors came, they saw lettuce and cabbages. But the pump broke down after seven or eight months, and the project collapsed; they didn't farm the second year, nor the third year, nor the fourth. The garden died. No project has ever succeeded in Bakel.

Adrian Adams

Soon afterwards, a journalist from *Le Monde* visited us briefly, in the company of the representative of OMI in Senegal.

Jaabe So

In 1990 I applied to the Baalu *communauté rurale* for a paper acknowledging my claim to the land Famaxa Sisoxo gave me at Seegankaani, that I had been farming since 1965. The Kuŋani *groupement* applied at the same time for a title to the land at Sanba-Salu which it had been farming since 1975. Over a year later, we had received no answer; we found out that of all the people who had applied for land grants, only *conseillers* and their relatives and friends had been successful.

Less than a year later, a still more serious matter arose: our villages' right to their *jeeri* farmlands was called into question. The officers of the Federation sent a letter to the President of Senegal, Abdou Diouf, with copies to all officials concerned:

We the undersigned, officers of the Bakel Peasant Farmers' Fderation, have the honour of writing to draw your attention to a serious problem which has arisen in the *département* of Bakel, between the *communauté rurale* of Balu and the *communauté rurale* of Gaabu.

Shortly before the start of the last rainy season, some inhabitants of Kuŋani (a riverside village within the Baalu *communauté rurale*) who had gone to clear their usual farmlands, came into conflict with some inhabitants of Guñan, an inland village belonging to the Gaabu *communauté rurale*, who claimed that the land in question was now theirs by right.

When representatives of the two *communautés rurales* met in Guñan to discuss the matter, the representatives of the Baalu *communauté rurale* were amazed to see their Gaabu colleagues take out an official-looking document and announce that this document, drawn up by them at a meeting in Tambacounda (a meeting to which it seems no representatives of the Baalu

> *Tomorrow you risk invasion by multitudes of Africans, who driven by extreme poverty will surge into the countries of the North, wave after wave. . . . They will be like the hordes you experienced during the Middle Ages. . . . It is in your interest to help Africa develop. Your government aid and business concerns should help keep our human masses here, to keep them from assailing your continent.*
>
> (President Abdou Diouf, interviewed by Le Figaro, 3 June, 1991)
>
> *Nearly 70 percent of Black Africans settled in France belong to the Soninké ethnic group. Eternal immigrants, they will always emigrate. To realize this, one need only visit the arid lands of their birth and despair. Let Europe lock its borders and lull itself with illusions. They will keep on leaving. They have no choice. The River region . . . is meant to become the country's rice-basket in the next few years. . . . This implies that Senegal will give up a pricing policy that even the late Soviet Union renounced ages ago. . . . In the meantime, emigration remains the Valley's main resource. . . .*
>
> *With a few exceptions, attempts to return home more often than not end in failure. . . . As the Minister for Emigrants remarks, in any case they haven't much of a chance: 'You can't expect people who've spent twenty years sweeping the streets of Paris, to set up their own businesses.' . . .*
>
> *The land of their ancestors, the proud kingdom of Ghana, is now no more than a sterile moor peopled with ghostly baobabs and thorn-bushes. Their country is exhausted. The horizon is blocked. Banished from everywhere, the River people are condemned, probably for a long time, to exile.*
>
> (Article by Bertrand Le Gendre in Le Monde, 22 April, 1992)

communauté rurale were invited), restricts the territory of the Baalu *communauté rurale* to a one-kilometre-wide strip along the river, the rest being assigned to the Gaabu *communauté rurale*.

It is our duty, as leading members of an association of peasant farmer *groupements*, several of whom are based in villages of the Baalu *communauté rurale*, solemnly to advise Your Excellency that this decision is unacceptable. The territory of the riverside villages within the Baalu *communauté rurale*, namely Kuŋani, Golmi, Yafera, Arundu, which have been in existence for over two centuries, has always included farmlands several kilometres inland from the river. The enclosed map will show you that land farmed by the inhabitants of Kuŋani, in particular at Papata, Samba-Gawlo, Lugere, Jaabalu, Dogo, is three, four, and even five or six kilometres inland from the river; and this map does not show Gece, eight kilometres inland. . . . In addition to these lands, communal lands held in trust by the chief of Kuŋani, there are flood-recession farmlands belonging to specific families in Kuŋani which lie more than one kilometre inland from the river.

What is true of Kuŋani is also true of Golomi, Yafera, Arundu. Their farmlands are under the authority of their respective chiefs, who have never denied inhabitants of any other village the right to farm there, if they asked. That remains true today. But thus forcibly to deprive the riverside villages of an essential part of their farming lands, with the evident aim of seizing hold of irrigable land for the Gaabu *communauté rurale*, is very dangerous. It could set off an unending series of conflicts between people living by the river, and their brothers inland.

The *conseils ruraux* here do not function as they should. We have already had occasion to protest against certain abuses in the land grant system, without any result to date. The purpose of our letter is to ask Your Excellency to intervene as a matter of urgency to set things right by having the decision supposedly taken in Tambacounda, declared null and void. The problem of access to irrigable land can only be resolved by open negotiation between all parties; not by force or fraud.

The only reply we received was in April 1992, from Hamidu Saada Timmera, the president of the Baalu *conseil rural*:

1º/ ... In accordance with the provisions of Law 64–46 of June 17, 1964, the *conseil rural* alone is entitled to make grants of land belonging to the *domaine national*. Consequently, no village chief has any right over land belonging to the *domaine national*, as you claim in your letter.

2º/ If you are unaware of the existence of the *conseil rural*, how can you expect to receive grants of *domaine national* land when you already have several hectares that are not being used. (Note: I must ask you to refer to the terms of the above-mentioned law.)

3º/ You are entitled to write to the President of the Republic, but it would be a good idea occasionally to seek information first from those immediately in authority over you (*autorités de tutelle*).

Jaabe So

On a visit here, Christian Aid's representative suggested to me that I resign my position as president of the Federation, in order to devote myself to organizing peasant farmers all along the River. I said I couldn't do that; to prove my point, I took him from Arundu to Gande, holding meetings in each place. This showed him that the Federation was still faced with many difficulties, and people would not agree to my leaving. All the same, he asked me to consider taking on the work as an organizer. I said I couldn't, I was too tired. But in the end, I agreed.

> The Bakel area differs from other areas along the Senegal River:
> - There is higher rainfall, and rain-fed sorghum remains an important crop. Flood-recession farming is less important than further downstream.
> - The traditional Soninke extended family is still the most important socio-economic unit.
> - Most irrigated farming *groupements* in the area belong to the Federation of Organized Peasant Farmers of Bakel.
>
> There was a divergence of purpose between the two organizations which have together determined the development of irrigation in the area: the Federation supported the community, the integration of women, and developing irrigation to suit the needs of peasant farmers; SAED tended to divide the perimeter among individual men, and supported growing rice as a cash-crop, on credit. . . .
> Individualization, as it has taken place in the Bakel area, appears not as a progressive, but rather a regressive development, indicating a lack of confidence in irrigated farming.
>
> (G. E. Feenstra, Gestion de l'eau, November 1991, 'General Conclusions')

Adrian Adams

In January 1991 Christian Aid sent the Federation funds in support of the first year of its proposed programme of promoting contacts between peasant farmers' associations throughout the Valley.

We called on the Delegate of the European Communities Commission in Senegal, Mademoiselle Gabrielle von Brochowski, who advised us that as a result of the evaluation report that irrigated farming in the Bakel area could never be economically viable, USAID had decided to terminate its involvement there; and suggested we call on a Mr Delgado at USAID for confirmation. Mr Delgado pointed to a map of Senegal on the wall of his office, and drawing a line with his index finger from west to east through Diourbel, said: 'We're not funding anything north of that line.'

12

The End
(1991–1994)

Year Nineteen: 1991/1992

Jaabe So

Gande did not harvest anything: their field dried up. Gallaade and Mudeeri had a good harvest. Jawara did not cultivate their perimeter. In Yelingara, some people had a good harvest, and some not. Manayeli did not harvest anything. Tiyaabu did not attend the meeting. Bakel had a good harvest. In Kuŋani, some people had a good harvest, others not. Yafera did not cultivate their perimeter. In Arundu, some people had a good harvest, others not.

Adrian Adams

An enquiry into the irrigated, rain-fed, and flood-recession farming done by men and women members of the Kuŋani group in 1991/2, and the grain crops harvested by all Kuŋani households that same year (which was a year of moderately good rainfall), brought valuable information about current levels of food security and the part played by irrigated farming. This later became part of a study, produced with help from Christian Aid, which we used to seek funding for rebuilding the farming groups' irrigation systems.

Jaabe So

At the end of the rainy season, I visited the region we call Hayire. I held meetings in the towns of Denbankaani, Haadubere, Gurel Ja, Jella, Wawunde, Gumal, Jamgel, and Amadi-Wunaare. Then I travelled through Fuuta Toro, in the Middle Valley. I did not ask people to join our Federation, nor to make me their leader. I told them that their farming groups should unite and set up their own associations, in order to defend their interests.

I was made welcome everywhere. Farming groups around Denbankaani, Wawunde, and Amadi-Wunare said they would form

The End (1991–1994)

During the 1991 rainy season, Jaabe So farmed a third of an acre of irrigated rice and seven acres of irrigated maize at Samba-Salu. He paid 250,000 CFA francs in all, for irrigation expenses and ten fifty-kilo sacks of fertilizer; he harvested 500 kilos of rice and three tons eight hundred kilos of maize, and earned 2,225 francs from the sale of fresh ears of maize. They did not farm the *jeeri*. During the dry season he grew just under an acre of irrigated onion; he paid 30,000 CFA francs, for irrigation expenses and a sack of fertilizer, and harvested two and a half tons; if he had sold all of it, he would have earned 375,000 francs. The household also harvested 300 kilos of maize and about 200 kilos of sweet potatoes from their *falo* at Samba-Salu.

Jaabe So's household, while not a large one, receives many guests, seasonal workers, and apprentice mechanics; it consumes one and a half *muudu* of (hulled) rice a day, and three *muudu* of maize; they need one ton eight hundred kilos of rice and three and a half tons of maize a year. The household's harvest of irrigated maize represents fourteen months of its yearly needs; the harvest of rice, only two months. Thus the household is self-sufficient in maize; and lacks one ton two hundred kilos of rice, costing 160,000 francs to buy, which can be paid for from its profits from farming.

During the 1991 rainy season, Al-Haji Mpali Tanjigoora farmed half an acre of irrigated rice, a third of an acre of irrigated maize, and three-quarters of an acre of irrigated sorghum at Sanba-Salu. He paid 48,000 CFA francs in all, for irrigation expenses and twenty-five kilos of fertilizer; he harvested 1400 kilos of rice, 400 kilos of maize, and 600 kilos of sorghum. During the dry season he grew a half-plot of irrigated onion; he paid 2,200 francs, for irrigation expenses and two kilos of fertilizer, and harvested 400 kilos; if he had sold all of it, he would have earned 60,000 francs. He also harvested a hundred kilos of maize, 600 kilos of sweet potatoes and a fair quantity of bitter eggplant from his *falo*.

The household to which Al-Haji Mpali belongs, headed by Al-Haji MaJaaxon Ba Tanjigoora, is a very large one, consuming three *muudu* of (hulled) rice and five *muudu* of sorghum or maize a day; as thirty *muudu* amount to one hundred kilos, it needs three tons six hundred kilos of rice a year, and six tons of sorghum or maize. So his irrigated crops represent three months of the household's needs in rice, and two months of its needs in sorghum or maize. Even though the 1991 rains were good, the household harvested only two tons of sorghum, or four months' worth, from its rain-fed *jeeri* field at Garsingide. Thus, even if Al-Haji Mpali were to deposit his entire irrigated crop in the household granary, the household would not be self-sufficient in food, as it would have produced only six months' worth of sorghum or maize, and three months' worth of rice.

Even in a good year, the household lacks three tons of sorghum or maize and two tons seven hundred kilos of rice, which will have cost over 500,000 CFA francs to purchase. In a year when the rains fail, like 1992 when they harvested only a few hundred kilos, they will have had to purchase nearly 700,000 CFA francs' worth of grain. Al-Haji Mpali's hard work in irrigated farming cannot in any way ensure food security for the household, which is almost entirely dependent on household members working abroad.

Another group member, Manju Jeba Tanjigoora, grew half an acre of irrigated maize during the 1991 rainy season; he paid 18,000 francs for irrigation expenses and sixteen kilos of fertilizer, and harvested 600 kilos. During the dry season, he grew over a third of an acre of irrigated onion; he paid 12,500 francs, for irrigation expenses and twenty-five kilos of fertilizer, and harvested one ton two hundred kilos, which if sold would yield 180,000 francs. Manju Jeba's household in Tanjankunda, headed by his elder brother Sanba Jeba, is a large one, consuming one and a half *muudu* of (hulled) rice a day, and four *muudu* of sorghum or maize; they need one ton eight hundred kilos of rice and four tons eight hundred kilos of sorghum or maize a year. Manju Jeba's harvest of irrigated maize represents one and a half months of the household's yearly needs. The 1991 rains were good: the household harvested five tons six hundred kilos of sorghum from its rain-fed *jeeri* field at Gece, ten kilometres from Kuŋani. This represents fourteen months of their requirements; they also harvested half-a-ton of *falo* maize, which makes it fifteen months. But in a year of poor rainfall, like nine of the past ten years, the household would be far from self-sufficient. Thus in 1992, they harvested only three hundred kilos of rain-fed sorghum, less than a month's worth; the household lacked nine and a half months of sorghum or maize, three tons eight hundred kilos, which will have cost over 250,000 thousand CFA francs to buy, not to mention the one ton eight hundred kilos of rice, costing nearly 250,000. So Manju Jeba's irrigated crop, while qualitatively important for the household, cannot at present ensure its subsistence if rain-fed crops fail.

Manju Jeba's wife Xujeeji Tanjigoora grew one plot of irrigated rice during the 1991 rainy season; she paid 3,300 francs, for irrigation expenses and eight kilos of fertilizer, and harvested 250 kilos. She also grew rain-fed groundnuts and okra at Suxangide, three kilometres from Kuŋani, and harvested 400 kilos of groundnuts and 200 kilos of okra. During the dry season she grew two plots of irrigated onion, paying 4,800 francs for irrigation expenses and eight kilos of fertilizer, and harvested 450 kilos. Three other women of the household also belong to the farming group, and harvested similar irrigated and *jeeri* crops. These crops mean extra food for their children, better food for the household as a whole, and a bit of money of their own from the sale of onions and rice.

During the 1991 rainy season, the women of the Kuŋani farming group grew three and a half acres of irrigated rice; they harvested six tons two hundred kilos in all. They also did much of the work on the five-acre collective field, which yielded seven tons of paddy; in acknowledgement of the fact that irrigation is carried out by the men. Like most of the women of Kuŋani, all these women also farmed rain-fed groundnuts. The rains were good; the group's women harvested fourteen tons eight hundred kilos of groundnuts in all. During the dry season, they each grew one or two plots of irrigated onion, two and a half acres in all, harvesting a total of sixteen tons; also a one-acre collective field which yielded two tons of onions, sold to provide funds for the women's own *caisse*. They also took part in work on a one-acre collective field for the benefit of the group as a whole.

During the 1991 rainy season, the men of the Kuŋani farming group farmed thirteen acres of grain under irrigation; that being the area that can be farmed without unacceptable expense and risk in the present state of the irrigation system. They harvested fourteen tons one hundred kilos in all, eleven tons six hundred kilos of maize or sorghum and two and a half tons of rice. The households to which these men belong need at least twenty-four tons of rice and forty tons of maize or sorghum a year. Their irrigated harvest represents one month's supply of rice and three and a half months of maize or sorghum.

In 1991 the men of the farming group harvested twenty-three tons of rain-fed grain and two tons of flood-recession maize, representing seven and a half months' supply. Their entire harvest thus supplies eleven months of their yearly needs in maize or sorghum, and one month of rice. So their households' food deficit for 1991/2 is four tons three hundred kilos of maize or sorghum and twenty two and a half tons of rice. Purchasing this quantity of grain would cost at least 3,300,000 CFA francs. Let it be remembered that the rains of 1991 were exceptionally good, or at any rate the best for the last fifteen years. As an example of a poor rainy season, one can cite 1982, when the inhabitants of Kuŋani made written note of each household's rain-fed harvest as compared to that of the previous year, in order to ask for food aid. The total crop declared that year by the households of the men now belonging to the farming group was ten and a half tons of maize or sorghum, compared to twenty-seven and a half tons for the previous year; amounting to three months of their yearly subsistence. Even taking into account their irrigated crop, they would have to buy at least 5,500,000 francs worth of grain.

To ask whether irrigation is 'profitable', does not in itself make much sense. Profitable for whom? From the point of view of the group as a whole, their paddy rice cost them 43,000 francs a ton to produce; purchasing the equivalent, 600 kilos of hulled rice, would cost 81,000 francs. Their maize cost them 30,000 francs a ton to produce; purchasing the equivalent would cost at least 75,000 francs. Their sorghum cost them 40,000 francs a ton to produce; purchasing the equivalent would cost at least 60,000 francs. From the treasurer's point of view, the group's accounts for 1991/2 were in the black: total receipts, from members' dues, sale of seven tons of paddy from the collective field, and sale of onions from the collective field, amounted to 1,416,000 francs, and total expenditure, including thirteen drums of fuel, twelve sacks of fertilizer for the collective field, pump maintenance and repairs, and onion seed for the collective field, amounted to 914,000 francs.

Irrigated farming is profitable in the short term for the three farmers above, in the sense that a kilo of maize, rice, or onion costs them less to produce under irrigation, than to buy. But each of them, at present, could give it up and rely on money orders from family members working abroad, supplemented by rain-fed farming, or a pension from the merchant marine. They have persevered with irrigation in order to keep alive its promise for the community, knowing that neither rain-fed crops nor income from emigration can provide security for the future.

True, this way of seeing things is valid only in the short term, for it does not take into account pump amortization; deducting contributions to an amortization fund would push accounts into the red, if only because of the smallness of the area under cultivation. But saying that leads nowhere. For the fact is that no one has ever been concerned with making these perimeters profitable for peasant farmers; that would have required, for a start, constructing proper irrigation systems. For SAED, the perimeters were profitable from the moment they existed on paper. The peasant farmers have inherited a burden for which they are not responsible. The just and useful question to ask, is not whether irrigated farming is at present profitable, but whether it can be.

(Federation paper, 'Irrigated Farming and développement paysan')

---

federations. In January a delegation from the Denbankaani area came to Kuŋani to find out how they should go about it. We visited each other several times more; they set up their association and prepared the papers needed to apply for official recognition. Throughout the year, I continued my travels along the River.

After all the building we'd done with the money left from Oxfam America funds for the fuel tanks and workshop, there was still some money left. They agreed to let us use it to buy a rice thresher and a thresher for sorghum and maize.

PRIORITIES FOR SOLIDARITY

Peasant farmers' organizations

Peasant farmers are the majority of the Senegalese population, and constitute the poorest and weakest groups. Support to peasant farmer institutions, aiming to strengthen peasant farmers as a group and protect their environment, is a high point of the priorities set by Christian Aid and its partners. . . . Authentic Senegalese peasant farmers' associations can still emerge. One would hope that they would constitute the embryo of a peasant farmers' network or federation, representing their socio-economic interests and acting as a pressure group.

By Geographical Location

The Senegal River Valley

. . . Christian Aid is taking part in efforts to help peasant farmers face up to the challenge [of River development plans] by providing resources and knowledge. More than material support may be needed for effective solidarity, if the Senegalese government is to be called upon:
(a) To define clearly the long-term aims and strategy of après-barrage policy, and the part to be played by peasant farmers.
(b) To give necessary support, in particular time for farmers to adapt to modern irrigated farming.
Of course, it is up to our partners to approach the government. Our role could be—with others—to bring pressure to bear on Senegal's bilateral and multilateral donors, to get them to adopt a policy of dialogue with the government on the above points. In view of the urgency of the problem, this should be considered as a short-term plan of action with long-term implications. There may already be too little time to launch this process. . . .

(Christian Aid policy document on Senegal, 1989)

Adrian Adams

The Federation's report to ICCO on support for women's market gardening noted that its future expansion depends on solving two problems: the poor quality of existing irrigation systems, which means unacceptably high risks and costs, and marketing difficulties.

Towards the end of 1991 the literacy programme produced several Soninke-language publications: *suras* from the Koran, the first issue of a magazine called *fanqanne*, and a book on River fish and fishing, entirely made up of contributions by farming group members. It was

The Federation was officially recognized in July 1985 ... but this can be seen as a hollow victory: not an acknowledgement that the Federation's struggle was well founded, just an acknowledgement that since 'administrative development' had fallen into disfavour for quite different reasons, the Federation's struggle—essentially, for peasant farmers to be considered as citizens by the State—was no longer of much consequence.

A second generation of perimeters has appeared, which has no links with the Federation. Federation farming groups have lost some of their membership; people who rejected SAED's priorities, but did not believe anything else was possible, gave up irrigated farming saying it was not for poor people. Those who as founders and leading members of the initial groups, persevered with irrigated farming, did so in order to retain a hold on irrigable land and keep open peasant farmers' prospects in that field. ...

From 1985 to the present, the Federation has organized the supply of goods (fuel, fertilizer, seed) and services (pump maintenance and repairs, tractors, literacy and training), all managed by farming group members. ...

However, the Federation's achievements have not been acknowledged, still less supported, by the authorities; since 1985, just as before, neither SAED nor USAID have treated the Federation as a rightful partner. This is surprising, as the Federation's activities are in line with the New Agricultural Policy. But the Federation seems to be just as anomalous in the era of the private sector, as it was in the era of State control.

The New Agricultural Policy recommends both 'giving responsibility to peasant farmers' and 'promoting the private sector'. That implies that 'peasant farmers' and 'the private sector' are two separate and distinct entities. But what does 'the private sector' consist of? For the Bakel area, if one considers what actors have come to the fore since the 'withdrawal of the State' was announced, what does one find? Those for whom irrigation systems (*aménagements*) have been built, are never entrepreneurs investing their own capital, but government officials or politicians setting up in business with subsidies from public funds.

SAED has remained present to oversee *aménagements*. USAID remained present for a time on the sidelines, having sent to Bakel a team of consultants meant to prepare future *aménagements*, who did nothing and left when USAID decided not to put any more funds into River development. Officials, especially the French, spoke of a hypothetical 'reintegration of migrant workers', and a few 'returned migrants' attempted to set up pilot ventures on their own. In such cases as well, activities touted as 'private sector' were in fact subsidized with public funds.

The End (1991–1994)

It is high time to assert that in a country like ours, the basis of the private sector is family farming: the subsistence and cash crops, rain-fed and irrigated, grown by men and women peasant farmers. As a peasant farmers' association created to support family farming, the Federation must take its rightful place. Now as in the past, only peasant farmers' associations like the Federation can give meaning to words like 'food security', 'income creation', 'making farmers responsible', 'promoting the private sector'; they alone can reconcile them, and protect the community's way of life while encouraging their members' initiatives.

(Federation paper, 'Irrigated Farming and développement paysan')

---

decided to revise and expand this book for wider publication, to do justice to the wealth of material.

We wrote to Abiy Hailu of Christian Aid in October 1991:

For the past few years, the poor condition of their irrigated perimeters has created serious problems for our farming groups. This preoccupies us greatly, because only good irrigation systems can enable them to benefit from all the Federation has achieved up to now....

When we last spoke of this with Mademoiselle Gabrielle von Brochowski, then Delegate of the European Economic Community in Dakar, it was agreed that we would draw up a funding request along the lines required by FED Microréalisations.... Mr Greiling, Counsellor at the EEC Delegation in Dakar, suggested that as Lomé VI funds for Microréalisations are exhausted, and Lomé VII funds are not yet available, we should address our project to a European NGO, which would then submit it to Brussels....

We immediately thought of Christian Aid, as having among our NGO partners the most coherent vision of the problems of the Valley and the role that can be played by an association like ours....

In January 1992 Christian Aid sent the Federation funding in support of the literacy programme. They had previously agreed to make part of it available in the UK, which made it possible for me to purchase a computer and printer. In February 1992 Christian Aid sent the Federation funds in support of the second year of their programme for promoting contacts between River peasant farmers' associations. They also sent funds for a feasibility study for a programme of rehabilitation of irrigated perimeters.

Jaabe So

At the Christian Aid partners' meeting in Ndioum, November 1991, there were not many peasant farmers' associations; mostly people from NGOs. Among them was Mazide Ndiaye, of RADI [Réseau Africain

> For RADI, the development of our continent must be integrated, and centred on man. All strategies and policies for national development must rest upon extensive participation by the population.
>
> USE, [Union pour la Solidarité et l'Entraide], whose motto is SERVICE, seeks to participate as far as possible to establishing and developing solidarity and mutual assistance in Senegal, in Africa and in the world.
>
> In rural areas, USE seeks to favour the emergence of a community-centred, grassroots form of development, based on the responsibilisation of peasant farmers.
>
> (Presentation documents circulated at the Christian Aid meeting)

pour le Développement Intégré], who used to be with OFADEC. He suggested that River land be divided into three parts: one for River people, one for people from elsewhere in Senegal, one for agribusiness.

Representatives of the Federation and of USE, a Senegalese NGO active in the lower valley through the Programme Intégré de Podor, decided to work together to organize a meeting of River peasant farmers' associations, in the hope of agreeing on a statement voicing their shared preoccupations.

Adrian Adams

Julius E. Coles, Director of USAID in Senegal, Racine Kane from the Cellule Après-Barrages (CAB, attached to the Ministry of Planning, and responsible for overall planning of River development) and Muneera Salem-Murdock, research director of the Senegal River Basin Monitoring Activity (SRBMA, funded by USAID), visited Kuŋani in February 1992. We were informed on this occasion that USAID had set up a project to support non-governmental organizations, through which it would be able to assist peasant farmers' associations such as the Federation. Racine Kane undertook to let us have a copy of the summary version of the CAB's Master Plan.

We had already received a copy of SRBMA's final report, which had single-handedly reintroduced into current debate the idea that releasing a yearly artificial flood from the Manantali dam, so as to ensure the preservation of flood-recession farming, reproduction of River fish, and dry-season fodder for cattle (as against releasing only enough water to maintain a sufficient level for irrigation, and retaining the rest for hydro-electric energy) was a valid proposition even in purely eco-

The End (1991–1994)

Towards the end of the 1980s, all these failures and fears for the future led to a change of direction:
- first there was the New Agricultural Policy, which since 1984 seeks to disengage the State from the productive process as such, and to responsibilize the producers themselves, the most concrete token of this being the gradual withdrawal and restructuring of SAED;
- then there was the encouragement of private investors to invest in irrigated farming, with a sort of 'new frontier' opened to them in the north of the country, which raises the double problem of land tenure and respect of minimum construction norms;
- finally, there was a renewal of ideas on the type of development chosen, the idea of maximizing irrigated surfaces yielding to that of an integrated and harmonious form of development, achieving the best possible compromise between social considerations (self-sufficiency in food for the population), economic considerations (profitable returns on investments) and ecological considerations (restoration and protection of the environment).

This is precisely the context of the 'Master Plan' which aims to define a development strategy for the left bank for the next 25 years.

> (Introduction to the Summary Version of the Plan Directeur de Développement Intégré pour la Rive Gauche de la Vallée du Fleuve Sénégal, Ministry of Planning of the Republic of Senegal, UNDP, IBRD, October 1991)

Development initiatives in the River basin continue to concentrate on irrigation alone, even though it has rarely fulfilled its production or income-generating goals in the past. This poor performance may be attributed to its (generally unspecified) manpower and capital requirements, constituting too heavy a burden for many households, which use most of their capital to purchase food, and their manpower on a diversified range of productive activities designed to minimize risks. A comparison of the net rate of profitability of the manpower and capital invested per unit of land area, indicates that in years of adequate rainfall and flooding, the factors of production in shortest supply—manpower and capital—yielded a better return through flood-recession farming than through irrigated farming. A worthwhile development strategy should put irrigation in its place, as one and only one of the elements of a complex production system. . . .

Development of the Valley makes sense only if it is aimed at the majority, by creating new local opportunities to halt the cycle of impoverishment, ensure food security and multiply sources of income.

The most urgent of these goals remains food security, which can only be ensured in the immediate future by rapidly rehabilitating the irrigated

> perimeters, preserving and improving flood-recession farming, protecting rain-fed farming against pests and increasing the productivity of animal husbandry.
>> (Final Report of the Senegal River Basin Monitoring Activity, Institute for Development Anthropology, Binghamton, NY, May 1991)
>
> The Project to Support Non-Governmental Organizations is an eight-year project, funded by USAID to the amount of 15 million dollars. It started at the end of 1991.
> The Project is executed by a co-ordinating structure, the NGO Support Unit, set up and managed by an American NGO, New Transcentury Foundation. . . .
> The aim of the Project is to enable national non-governmental organizations and community groups to plan, conceive, and carry out viable development activities, alone or in collaboration with American NGOs. The idea is to transfer directly to communities, through NGOs, the resources necessary to achieve their goals. . . .
>> (Information sheet left after a visit to Kuŋani)

nomic terms. Unfortunately, SRBMA came to an end, and Muneera Salem-Murdock left Senegal in 1993.

From February to April 1992, the Federation commissioned topographical surveys of the irrigated perimeters of the six farming groups included in the first phase of rehabilitation: Yelingara, Manayeli, Tiyaabu, Bakel-Gassambilaqe, Kuŋani (both Sanba-Salu and Seegankaani), and Arundu.

Jaabe So

After the meeting organized by Christian Aid in Ndioum, we agreed to work with USE to draw up a statement of the needs and fears of River people, that would be like a flag for people to follow. Once that had been done, we organized a meeting of River associations to discuss it.

The meeting was held at Ndioum in April 1992. It was attended by twelve associations, including the newly formed Fédération des Paysans Organisés de Dembankané, who agreed on a document in Pulaar and Soninke.

Adrian Adams

After the return to power in France of a right-wing government, rejection of Arab and Black African immigrants became a widely ac-

APPEAL BY PEASANT FARMERS' ASSOCIATIONS OF THE SENEGAL RIVER VALLEY, MEETING IN NDIOUM ON 22 AND 23 APRIL 1992

Within living memory, the inhabitants of the Senegal River Valley have been able to make a living there.

Rain-fed and floodland farming ensured their subsistence, with herding and fishing in the river and ponds. Trees gave wood for building and cooking, as well as fruits and medicines.

Rain-fed farmlands were left to lie fallow, and the river's spate brought new soil to the floodlands: the land's value was preserved.

Just as people respected their environment, they respected one another. Those who went to work elsewhere, did so to help their families live better at home. They knew that the life they left behind had a solid base.

Older people knew that time, when one could trust in life. But younger people have known only a time when everyone fears tomorrow.

The lack of rain and flood has for years threatened the livelihood of all those who depend on the river. In spite of their hard work, those who stay at home harvest very little, sometimes nothing. Cattle die. Trees die. Fish have disappeared. The land is exhausted. Those who emigrate must send their families the wherewithal to survive.

As for plans to develop the river, they may be our death sentence, the final loss of hope: whether based, as yesterday, on State control, or as today on short-term profit, these plans exclude us. River peasant farmers have no place in River development schemes.

Faced will this danger, we who have signed this document affirm:

• that the main resource of a country like ours lies in the skills and endurance of the men and women who inhabit it: of us all;

• that the real 'private sector', the only possible foundation for development in accordance with professed aims (food security for River inhabitants, a chance to earn a living at home, protection of the environment by harmonious development of the area's full potential) is peasant family farming;

• that this peasant family farming can only survive and develop through forms of community solidarity organized by the peasant farmers themselves.

1. *We ask all those, peasant farmers, herders, fishermen and their families, who make their livelihood from the River, to create*, to join wherever they already exist, to strengthen if they belong to them already, genuine organizations, without partisan or sectarian affiliations, respecting the rules of internal democracy and financial openness, and in particular:

- strong *farming groups* at village level, to cope with problems that cannot be dealt with by families on their own: ...
- *peasant farmers' federations* between villages, to cope with problems which can only be dealt with at the level of one or several *communautés rurales*: ...
- *active co-operation between peasant farmers' federations* of the River Valley, in the hope of one day forming a peasant farmers' union to deal with problems which can only be solved at the national and in some cases international level: ...

2. *We ask the administrative authorities to acknowledge River peasant farmers' organizations as significant partners*, rather than seeking to bypass them as is all too often the case, and in particular:
- to give an appropriate legal status to peasant farmers' organizations, and simplify the procedure for gaining official recognition;
- to give genuine peasant farmers' organizations access to tax-free inputs and agricultural equipment: they should not have to seek NGO status, which denatures their role by treating them like organizations intervening from outside;
- in co-operation with peasant farmers' organizations, to promote a pricing and import policy favourable to local producers of cereals, onions, tomatoes, and other vegetables;
- in co-operation with peasant farmers' organizations, to regulate the artificial flood in such a way as to favour flood-recession farming and the reproduction of River fish;
- in co-operation with peasant farmers' organizations, to evolve a land grant and development policy that gives priority, first to the present and future needs of River inhabitants, then to the present and future needs of the inhabitants of the rest of Senegal, and which takes into account all possibilities for developing the land, not just irrigation;
- in co-operation with peasant farmers' organizations, to seek appropriate forms of funding for *aménagement*, other than recourse to *crédit agricole* in its present form;
- to take on responsibility for public health, especially for illnesses linked with water, such as malaria and schistosomiasis;
- to enforce the laws on *conseils ruraux*, in particular those concerning the public's right to attend *conseil* meetings, the posting of minutes of meetings, land surveys, and procedures for making land grants;
- to give the peasant farmers', herders' and fishermen's associations of a *communauté rurale* the right, formerly enjoyed only by co-operatives, to elect a number of *conseillers ruraux* equivalent to one-third of the number of those elected by universal suffrage;
- to allow independent candidates in local elections.

3. *We ask NGOs intervening in the River Valley to give direct support to peasant farmers' organizations*, and in particular:
• to promote self-reliance on the part of peasant farmers' organizations: by supporting production, with systems for supplying inputs and services which can thereafter be self-financing; by making available information (on present activities and future plans of River development schemes) and training;
• to avoid subsidizing the private commercial sector, which enforces dependency on peasant farmers, without itself being economically viable;
• to judge peasant farmers' organizations by their representativity, internal democracy and financial openness, rather than their conformity with current Western discourse on development;
• to be wary of giving automatic priority to women and youths, as this can also be a way of evading important questions about representativity and power;
• not to go along with European governments by treating 'migrant workers' like a population apart.

---

ceptable political stance. Many Soninke men working in France, since earlier restrictions on immigration meant they could no longer look forward to being replaced by a brother or son, had by then had their wives join them; not only single men living in hostels, but whole families, were now at stake. Two of Jaabe So's daughters, Juumu and Ya, had been living in France with their husbands, in Montreuil and Epinay-sous-Sénart, since the early 1980s.

Year Twenty: 1992/1993

Jaabe So
Gilli ken siine, in 1989 we had applied for the Federation to be granted NGO status, which would entitle us to duty-free equipment. Since then, whenever we were in Dakar, we would contact Cheikh Amar at Community Development, to find out what was happening to our dossier. First he told us that all applications had been frozen until the passage of a new law on NGOs; then he told us that the dossier had been sent to the Governor for comment . . . All this when small groups with only a few members, in Casamance for instance, were being granted NGO status only months after asking for it.

When Racine Kane of the Cellule Après-Barrage visited Kuŋani, he undertook to support our application. He wrote shortly thereafter that the previous application, referred from Dakar to the Governor of Tambacounda, had never reappeared; a new application would be necessary, which he would follow up himself. A new application was duly made.

In December 1992 we heard that the Caisse Française de Développement was funding another large market gardening project in the Bakel area (the sum mentioned was six hundred and fifty million CFA francs). To try and stop the money vanishing this time, they asked GRDR to supervise the project. They had been in Bakel for several years, supervising various returned migrants' projects, mostly in Mali and Mauritania, but had no relations with the Federation.

Benoît Thierry, the GRDR co-ordinator in Dakar, said they wanted peasant farmers' associations like the Federation to be represented on the project. He acknowledged that local politicians would probably try and take control, which would put the GRDR in an uncomfortable position.

In March 1993 Roger Briend, appointed by the GRDR as director of the Projet Horticole de Bakel (Hortibak), came to ask me to help ensure that the project was not taken over by politicians, but benefited farmers. I agreed to meet him again to discuss the matter, and to attend a meeting the following week at the *préfecture*.

When I met him again, he told me that there were problems: he had been advised that the steering committee was to be based in Dakar, and made up mostly of officials from the Department of Agriculture and other government departments, with the Caisse Française de Développement, GRDR, and two representatives of farmers' associations.

GRDR had suggested that there also be a follow-up committee in Bakel, with project beneficiaries taking part, headed by the *préfet* rather than by the Agriculture Service. But Roger had just received a phone call from the Caisse Française, saying that the Minister, Gawusu, was very upset, and Roger must come to Dakar straight away.

A few days later, we heard that Gawusu had telephoned the *préfet* to instruct him not to hold any meeting to do with the project.

Later on, farmers' representatives from the Bakel area elected three delegates to the steering committee, including me. The Ministry rejected these delegates; we met again and re-elected the same people.

I sometimes wish the Caisse Française had simply put the money in Gawusu's bank account, and saved everyone a lot of trouble. There were all kinds of meetings, and I attended faithfully, but the project produced only paper. Things went on we weren't told about, and it seemed as if whatever the project was for, it wasn't for helping farmers. We were told it wasn't meant for irrigated perimeters, it wasn't meant for crops other than vegetables. If women involved in market gardening wanted their own pumps, so as not to depend on the men, they would have to pay most of the money themselves. And when we talked about the problem of transport for marketing vegetables, Roger suggested donkey carts.

I didn't take part in this for the Federation's benefit; we don't need their money. I took part to try and stop politicians misappropriating project funds and making a mockery of honest people's work. Either the project would benefit farmers, and that's not happened yet; or we would denounce it publicly.

Adrian Adams

The 1993 elections in Senegal were fiercely contested in urban areas, where opposition parties achieved high scores and challenged the results. Abdou Diouf was returned to power thanks to the rural vote.

The Federation's study on perimeter rehabilitation was drafted by January 1993. In February there was a meeting of the presidents of the six farming groups to be involved in the first phase of perimeter rehabilitation, along with Federation officers, to discuss its contents and in particular the levels of costs and amortization payments required to ensure economic viability. After discussion, agreement was reached. The study was completed, and sent to Christian Aid in March 1993.

In May 1993 there was a meeting of the Federation's executive board, made up of the Federation's officers and of the members of the Federation's four working commissions: Agriculture, Equipment, Outside Relations, and Literacy and Training. Since early 1991 such meetings had been held twice yearly, in May and November, to evaluate the past season, with a complete review of income and expenditure, and prepare the coming season. This made it possible to state that total aid received by the Federation, from the beginning (Oxfam funding for Soninke-language literacy, 1981) up to April 1993, including the full value of equipment, amounted to 194,500,000 CFA francs ($778,000 at the 1977 rate of exchange, $1 = 250 CFA francs), from Oxfam

MANANTALI AND DIAMA: THE DAMS OF HOPE

There were great celebrations last Thursday all along the Senegal River Valley, on the occasion of the inauguration of the Diama and Manantali dams. . . .

'This has been a magnificent day from all points of view, for it has enabled us to show the world what can be accomplished when three neighbouring States unite to carry out a well-thought-out policy, and I feel it is a lesson for Afro-pessimists', President Abdou Diouf stated on his return from Manantali and Diama. . . . 'When Africa builds, carries out projects like these, which will last for generations and release 375,000 hectares of irrigated land and 800 million kwh a year, not to mention all the other promising results, no one talks about it. And they'll still say that Africans are incapable, that they can't build anything. . . . After this day, I am convinced that Africa will win its struggle.' He also expressed satisfaction with the behaviour on this occasion of the populations the dams are to serve. 'We are servants and we want to serve our peoples', he added.

(Article in the Dakar newspaper Le Soleil, 2 November 1992)

MRD [Ministry of Rural Development] plans for Bakel are linked with the PASA [Plan d'Ajustement Structurel Agricole] for rice-growing areas. The goal for the Bakel area is to put 6,000 hectares under irrigated rice, with 'heavy' focus on the Falemé area.

(USAID report of a conversation with the Minister of Rural Development, Cheikh Cissokho, in March 1990)

My concern with development began over thirty years ago when I met Father Lebret, whom the government of Senegal had asked to lay down guide-lines for the first economic plan.

Since then, like many others, I have tried as best I could to follow in his footsteps in my thinking and action. Yet the poorest of the poor are now poorer than ever, and more numerous. Some of those I have met during the past thirty-two years have made me the gift of their friendship and affection. I have received a great deal from them, while injustice weighs upon them even more heavily than before. My hope that the world would take on Father Lebret's project for universal development now seems very remote indeed.

(Christian Bompard, in Foi et Développement, Feb. 1992)

Since rain-fed and flood-recession crops no longer suffice to feed families, and families can rely less and less on money from migrants abroad, members of Federation farming groups have only one option: to seize their chance, and make a success of irrigated farming. But the chance is not theirs alone. The groups have always been open to all; their present-day members are those who persisted in spite of all the difficulties. In defending their future, they are defending the future of all River peasant farmers.

The men often belong to a key generation, now between forty-five and seventy years old, which has experienced emigration and sought to place hope in something else. Many women would like to have more irrigated land to farm. This is not possible at present, because the perimeter is difficult to irrigate, and this is a source of dependency and tension with the men. The problem would be solved if the perimeter were to be correctly rebuilt, as the women could then carry out irrigation themselves.

The men and women who are now group members are not many. But many are those who will benefit from perimeter rehabilitation. First of all their households, which represent many people, as brothers and sons of brothers with their families produce and consume together; but also the other inhabitants of their towns. For present-day irrigated perimeters are far from occupying all irrigable land.

For Kuŋani itself, the *Plan Directeur d'Aménagement de la Rive Gauche* indicates in its Land Use and Distribution Map 1,650 further acres which could be brought under irrigation, 150 acres of which are suitable for rice, and 1,500 for mixed farming. And the other inhabitants of the area are in the same case as the group members: their rain-fed harvests are generally far from satisfying their basic subsistence needs.

Thus in 1991/2, the inhabitants of Kuŋani obtained from their rain-fed and flood-recession crops 165 tons of maize and sorghum. Their basic requirements can be estimated at 300 tons of grain a year. There was therefore that year a shortfall of about 135 tons, which at prices then current would cost over ten million CFA even in the most favourable circumstances, just for subsistence grain; and that in a good year.

The population of Kuŋani is growing rapidly. Chances of earning enough through emigration to support a family at home, are dwindling rapidly. The productive potential of Kuŋani's land represents the only hope for the youth of today of living a decent life. The same is true of all riverside villages in the Bakel area. The example of a successful combination of irrigation, rain-fed farming, and flood-recession farming favoured by an artificial spate, would be of great assistance to them.

But in order for irrigation to benefit the area's inhabitants (and also, by enabling them to define their needs, peasant farmers from other areas of

Senegal who might wish to come here) there will have to be a reversal of priorities in favour of peasant farming communities. Present trends in River development exclude them: at best, they are only temporarily reprieved.

Not that it is ever said clearly. At every stage of the road which has led to the present impasse, one meets with the same double language. Thus the law on National Domain tolerates the present use of rural *terroirs*—so long as 'the general interest' does not dictate otherwise. The OMVS programme gives priority to 'food self-sufficiency of River people'—so long as 'cash-flows are generated which lead to rapid growth'. The New Agricultural Policy proposes both to 'give responsibility to peasant farmers' and to 'promote the private sector'. Endlessly vacillating between what it is politically opportune to say, and what is enjoined by major funders, the government has let itself slide towards an irreversible state of affairs, by letting time act in favour of the powerful, which will be a catastrophe not only in political, but in economic terms.

It seems unlikely that large industrial companies will set up on the River, unless they are subsidized like the Compagnie Sucrière Sénégalaise. Redundant white-collar employees, retired government officials, unemployed degree-holders, politicians in search of a clientele, Islamic associations will be variously subsidized to become the new farmers. They will depend either on mechanization, or more durably on recourse to unwaged labour, *taalibo*, or poor peasants provided with advances to buy fuel and fertilizer, in exchange for half the harvest.

Peasant farmers who try to keep up will sink hopelessly into debt, in a context where everything is against them. The proliferation of GIE should not create any illusions: it is not enough to free peasant farmers from State management, to turn them into individual agricultural entrepreneurs. They can only succeed if organized.

If present trends continue, the result of River development without peasant communities will become apparent: a polluted river, an exhausted hinterland, perhaps a few cash-crop enclaves, but mostly abandoned perimeters; old people, women, and children living like refugees in their own villages; young men gone off in search of anything, anywhere. The refusal to subsidize peasant family farming will mean subsidizing misery: perpetual food aid, and funds to the authorities who maintain order.

A choice has to be made: either to base River development on peasant farmers, or to exclude them. It is known that large schemes, agrobusinesses, and crony bank loans for 'progressive farmers' benefit only a small minority. In the name of what should River communities accept a death sentence? On the contrary, giving priority to peasant family farm-

The End (1991–1994)

ing is in the general interest, making irrigation profitable and favouring the emergence of a genuine private sector.

Successful rehabilitation of the Federation's perimeters would be a step in this direction, showing in particular that well-managed aid can sustain future autonomy. This example would help other peasant farmer organizations on the River to seek perimeter rehabilitation, or to manage better the perimeters they have, in order to make them viable subsistence-based perimeters. This would in turn help peasant farmers from elsewhere have access to the River.

The introduction to the summary of the Master Plan for Integrated Development of the Left Bank of the River, says that 'so many failures and so many anxieties about the future, have brought about a certain reversal of trends', and especially 'a renewal of thinking about the chosen path of development, replacing the idea of maximum extension of irrigated surfaces with the idea of *an integrated and harmonious form of development*, achieving the best possible compromise between *social* (food self-sufficiency of the population), *economic* (return on capital invested), and *ecological* (restoration and preservation of the environment) imperatives.'

Only peasant farmer organizations can achieve such a synthesis; because for them, it is a matter of survival. Only they have an interest in refusing the plunder of their patrimony; only they have an interest in constructing systems which protect the land against erosion and improve its water retention, in regenerating the ponds where River fish breed. Only they have an interest in making irrigation part of the whole range of practices of peasant family farming, and in perpetuating irrigated perimeters by careful management, maintaining soil fertility, until such time as improved photo-voltaic technology may make it possible to dispense with imported fuel.

The most important thing for the future of the River is not irrigation, which is only one aspect of a livable future. The most important thing is that peasant farmers should accede to a degree of power. After so many years lurching from one failure to another in quest of new actors, there is no choice but to stake everything on *développement paysan*. Its victory is the only one where there will be room for all.

(Federation paper, 'Irrigated Farming and *développement paysan*')

America, Christian Aid, ICCO, Oxfam, War on Want, Mission Française, Comité Français Contre la Faim, Oxfam-Belgique, Agro-Action Allemande, and two Dakar-based consortia of NGOs, Collectif Inter-ONG de Lutte Anti-acridienne (CIONGLA), and Collectif pour la Sécurité Alimentaire à la Base (COSAB). Account of its use was

FEDERATION EXECUTIVE BOARD

Officers

President	Jaabe So (Kuŋani)
President (Women)	Goola Mayi Jallo (Manayeli)
Vice-President	Denba Bacili (Arundu)
Vice-President (Women)	Yaali Bacili (Gallaade)
Treasurer	Sanba Jarra (Manayeli)
Assistant Treasurer	Bullayi Kanute (Mudeeri)
Secretary	Adrian Adams So (Kuŋani)

Commissions

Agricultural	Manayeli Jallo (Manayeli)
	Denba Tarawore (Manayeli)
	Sili Tapa Bacili (Tiyaabu)
Equipment	Mpali Tanjigoora (Kuŋani)
	Denba Giise Saaxo (Jawara)
	Denba Goola Jallo (Manayeli)
Outside Relations	Bullayi Kanute
	Daramaanu Jarra (Bakel)
Literacy and Training	Adrian Adams So
	Mammadu L. Bacili (Arundu)
	Mammadu S. Bacili (Arundu)

faithfully rendered, and it was for the most part visible in buildings, fuel tanks, fertilizer stores, tractors, threshing-machines, and three cars (two pick-ups and a four-wheel drive), all in good working order; as in the competence and satisfaction of their users. What is lacking, is what the USAID Irrigation and Water Management Project so signally failed to provide: decently-built irrigation systems.

Total USAID funding for irrigated farming in the Bakel area from 1974 to 1990 (Pre-Project Funding, Bakel Small Irrigated Perimeters Project and Irrigation and Water Management I), amounted to $15,520,000 (3,880,000,000 CFA francs at the same rate of exchange as above), twenty times the total funding received by the Federation. Its only visible traces in 1993 were the pumps on the river, most of them ten years old or more, and the almost empty SAED buildings just outside Bakel, stripped of the panels for the solar pump that never worked. The irrigation systems built by SAED at the time of the Bakel Small Irrigated Perimeters Project, substandard from the outset, were by then largely unusable, or required considerable work and expense to operate. Some equipment and vehicles lay abandoned; some had been sold off. The rest of the money had been spent on salaries and working expenses: for SAED, for USAID-hired personnel, for US consultants; or remained unused, meant for irrigation systems that were never built.

According to the most recent and precise estimations, rehabilitation of all the Federation perimeters, about 1,500 acres, would cost about 720,000,000 CFA francs ($2,880,000 at the same rate of exchange as above), less than one-fifth of total USAID funding for irrigated farming in the Bakel area.

Year Twenty-One: 1993/1994

Jaabe So

In July 1993 the new SAED *Ingénieur-délégué* in Bakel visited me to say that SAED had hopes of new funding, and wanted to work with peasant farmers. I told him that the matter would have to be discussed by the Federation first. A few days later, I sent him an answer:

SAED has been in the Bakel area for nearly twenty years. During that time, it has always refused to work with the Federation; on the contrary, it has always sought to divide us. Faced with that refusal, the Federation did its best to function as an independent peasant farmers' association.

You say that you are going to have funds to construct new irrigation systems. Our farming groups, the first to start irrigated farming in the area, will have their share in any *aménagement* programme, whoever is in charge. Aside from that, if you really are going to return here, there is no reason why we should not be good neighbours. . . . But as for officially co-operating with you, forgive us, but the answer is no.

Adrian Adams

In September 1993 Julius E. Coles, Director of USAID Senegal, visited Bakel in the company of the US Ambassador to Senegal, Mark Johnson. He had advised us of his arrival, and a Federation delegation was on hand to meet him, with whom Mr Coles and the Ambassador conversed for about an hour. Mr Coles said in the course of conversation: 'We were wrong not to work with the Federation.'

In April 1994 Mr Coles left Senegal and USAID, to take up a post as Director of the International Affairs Center at Howard University.

Rural councils are elected, three-quarters by direct universal suffrage, and one-quarter by groups of an economic, social and cultural character, in particular co-operatives, groupements d'intérêt économique, sporting and cultural associations.
> (Article L 185, Law n° 92–16 of 7 February 1992, establishing the Electoral Code)

Candidates may also stand as independent candidates, unaffiliated to any political party.
> (Article R 57, Decree n° 92–267 of 15 February 1992)

Jaabe So

After the meeting in Ndioum, we held meetings every three months or so with the other associations that had signed the Ndioum declaration. The Ndioum declaration was sent to all levels of the administration, from the President of Senegal down to presidents of *communautés rurales*. There was no acknowledgement or reply. Later, we sent it out again; it was felt that perhaps the forthcoming election had distracted their attention. Once more, there was no acknowledgement or reply.

Adrian Adams

At the end of 1992, the Federation had been encouraged to apply for assistance to the USAID-funded Project to Support Non-Governmental Organizations, which, it was said, unlike USAID itself, would be able to work directly, not only with NGOs, but also with associations like the Federation. We asked for assistance in purchasing transport for our grain bank and motor repair workshop, essential components of the perimeter rehabilitation programme. In September of 1993, we were asked to provide further information, which we did. In July 1994 we received a letter informing us that the Project's activities were being suspended, for review in the light of USAID's new strategy.

Early in 1992, the Federation had made yet another application for NGO status. In January 1994, we received a letter from the Minister responsible, the Minister of Women, Children, and the Family, dated October 1993, saying: 'I have received your application. For statutory reasons (*des raisons d'ordre règlementaire*), I cannot accept it.'

We did not find this a satisfactory reply, and lodged yet another application. On 20 April 1994, we received a letter from the Minister granting Jaabe So an *audience* on 7 April. The letter was dated 10

March, but had been posted on 13 April. We wrote to explain why Jaabe So had not been able to attend.

On a subsequent trip to Dakar, we made enquiries at the Ministry, and learned that the most recent application had been mislaid; could we make another? We did, in June 1994.

Jaabe So

Manca Mjaayi was never a farmer, until he was made president of the Mudeeri *Communauté Rurale*. Then he granted himself 150 acres of someone else's land, and SAED put in canals. So he became a farmer. He went heavily into debt to Crédit Agricole, for equipment and fertilizer, and couldn't pay.

SAED began taking him to meetings in Saint-Louis. Then on 25 June 1994, he called a meeting at SAED in Bakel. I was the only person invited for the whole Federation; Daramaanu Jarra, the president of Bakel-Gassambilaqe, came along as well. Otherwise, the meeting was attended by political groups from Mudeeri, Jawara, and Bakel, that are not members of the Federation. Manca announced that Kuwait was going to give seven thousand million CFA francs for *aménagements*. Five thousand acres would be brought under irrigation in Bakel, with two thousand five hundred for Mudeeri alone; the work would be supervised by SAED.

He also announced that the Minister of Agriculture had asked him to organize federations of peasant farmers in the *départements* of Bakel and Matam. When Daramaanu and I pointed out that there was already a Federation for Bakel, representing eleven *groupements*, Manca said he had instructions from the Minister, and didn't need to take account of anyone else.

We wrote to the Minister on 30 June, reminding him that he was informed of the existence of our Federation, which had been officially recognized for the past nine years, and of our co-operation for the past two years with farmers' associations of the Middle Valley:

> We cannot believe, *Monsieur le Ministre*, that ten years after the proclamation of the New Agricultural Policy, and in our country's present grave crisis, you propose to disregard existing peasant farmers' associations and set up fake associations manipulated by the administration, by SAED and politicians. We hope to learn from you shortly that we have misunderstood your position. If it is true, however, that Monsieur Ndiaye's present manœuvers have the support of the administration, we will be compelled to denounce that as a scandal.

In July 1994 the twelve federations who had signed the Ndioum declaration met in Kuŋani to create an association uniting the River Valley's peasant farmers, the Mouvement des Acteurs de la Vallée du Fleuve Sénégal. I was elected president.

The Minister of Agriculture did not answer our letter, but sent it on to the *Président Directeur-Général* of SAED in Saint-Louis, who wrote to me on 3 August:

SAED is a part of the State and can only carry out the policy defined by the State. The major policy guidelines on Valley development are set forth in the Plan Directeur d'Aménagement de la Rive Gauche; as you must know, producers in general have mobilized in favour of its being entrusted to SAED, which they regard as their preferred partner. It was in the context of this partnership between SAED as a public service, and producers organized as they see fit because fully responsible, that SAED agreed to take part in the meeting organized by Mr Manthia Ndiaye, who was in fact simply responding to a request from a representative of producers from the Matam irrigated zone. You will be aware that above and beyond the union of Bakel peasant farmer organizations created at the 25 June meeting, Bakel and Matam producers have set up an even larger federation, covering both *départements*.

As for the Minister's instructions to Mr Manthia Ndiaye, there must have been a misunderstanding. According to my representatives as the meeting, the Minister merely advised representatives of producers to organize in order to take on responsibilities, rather than remain dispersed. I can assure you that there is no manipulation of associations by the administration, which is open to all.

People say Manca represents no one. But he's a politician; they don't need to represent anyone. You'd think the past twenty years hadn't happened. And I'm tired; I can't endure now what I endured in the past.

At the time of Independence, if we'd had our wits about us, we should have become independent as nations: Fuuta, Bundu, Gajaaga, each with their own ruler answerable to the people, just as the *mangu* and *sakko* could overturn a ruler's decision in days gone by. We should not have agreed to being all thrown together, so that a few could prey on all. If each nation had taken on responsibility for what concerned them directly, then all of Africa could have united.

If all the wealth that has entered Africa since Independence had been entrusted to nations, the people would have been able to supervise its use; it would not have been eaten. Whoever was given responsibility, whether *kome* or *hoore*, would have been held to account. As it is, once you've made me president or Minister, I can do what I like

The lordly Soninke, who had once and for centuries ruled the regnum of ancient Ghana with a pomp and circumstance that won the admiration even of Cordoba, that most civilized of medieval cities, might seem to have 'vanished from history.' But the chroniclers of the colonial partition, only 'yesterday', found the Soninke safely tucked away in the colony of Senegal, along the south bank of the river of that name; and that is where they live today. If the Soninke sometimes ponder the glories of their imperial past, there is little to show that it keeps them awake at night. If they have new aspirations to nation-statehood, they have not so far said anything on the subject.

Yet even a superficial acquaintance with the Soninke today tells one that they remain in their own thinking a distinctive people and culture, possessing a vigorous 'national consciousness,' whatever this may precisely mean, and a valid though little-written language. Their case is one among many, it seems to me, that underline the cultural misery of the whole nation-statist project. For postcolonial nation-statism in policy and rhetoric has preferred to talk down 'ethnic survivals' like the Soninke... as deplorably illegitimate and best forgotten, because, not having formed nation-states and therefore not being admitted to be nations, they have no right or reason to remain alive.

(Basil Davidson, The Black Man's Burden, *1992*)

---

with you. As it is, we are like hungry dogs, fighting over what little the *tubab* throw us. It has gone on for thirty years, and there is a colonialism ahead worse than the other, because now they can say: Do this, or we won't give you anything.

All the talk about democracy is a lie. There is no democracy now, nor will there ever be, unless it is based upon nations: upon our fathers' house, upon what we have.

Adrian Adams

We cannot reach conclusions. We do not know how the story will end, and that in itself is a victory; for we know how it should have ended. A thousand books have shown, since René Dumont began it, that development is a cruel hoax. This story has not been told to make it a thousand and one; but for its own sake, and to give our experience of what it is like to live through, why it continues, how it might be made to cease.

Those to whom this is done are consigned to oblivion, as if they did not exist. But there is no safety in that; they are too many. There will be no ease or mercy in life, unless, lies stripped away, silence broken, the claim is granted to a place of their own in our common present: a claim to land by the river.

Night has fallen. People have eaten together: steamed maize couscous made of last year's harvest at Sanba-Salu, still abundant in the granary, and a sauce of green cowpea leaves, the last of the small irrigated plot to survive the heat. Now they are resting and talking quietly, on the verandas facing each other across the narrow courtyard of beaten earth with its two neem trees.

There is a scent of rain in the air, a strong sweet scent of life renewed.

Two young men, Jaabe So's sons, are shelling maize for seed. The first of their irrigated fields is ready for sowing tomorrow.

A child of ten is looking through this book in Soninke. Turning to the first page, she begins to read: '*O na joppa ñiiñe ke*; let us begin with this land.'

Abbreviations

CIDR	Compagnie Internationale pour le Développement Rural
CNCAS	Caisse Nationale de Crédit Agricole
DAGAT	Direction de l'Administration Générale et des Collectivités Locales
ENDA	Environnement et Développement Tiers Monde
ENEA	Ecole Nationale d'Economie Appliquée
GIE	Groupements d'intérêt économique
GRDR	Groupe de Recherches et de Réalisations pour le Développement Rural dans le Tiers Monde
ICCO	Interchurch Organization for Development Co-operation
IFAN	Institut Fondamental d'Afrique Noire
MRD	Ministry for Rural Development
NGO	Non-Governmental Organization
NPA	Nouvelle Politique Agricole
OFADEC	Office pour le Développement Communautaire
OMVS	Organisation pour la Mise en Valeur du Fleuve Sénégal
ONCAD	Office National de Commercialisation de l'Arachide
SAED	Société d'Aménagement et d'Exploitation des terres du Delta du fleuve Sénégal
USAID	United States Agency for International Development
WARDA	West Africa Rice Development Association

Glossary

As a rough guide, Soninke can be pronounced like Spanish, with the following exceptions: *j*, *h*, *w* as in English; *x* like the Spanish *jota*; *q* like Arabic *qaf*, ŋ like *ng* in 'sing'.

anjoobe *Hydrocynus forskalii*
Baano the main Muslim feast-day of the year, the Arabic *Aid el Kebir*
balde *Heteroticus niloticus*
bantiŋe *Criodendron anfractuosum*
debegume town head
dere sauce made of cowpea leaves, thickened with groundnuts
fa *Zizyphus mauritania*
falo narrow strip of land along the inner bank, farmed as the river recedes
feela variety of sorghum
fonde porridge of granules of sorghum flour, eaten with sugar and sour milk
fosiwilisaana orator
fuñanŋe fonio
futo couscous of sorghum or maize
futuro sundown prayers
gajaba variety of sorghum
gese *Acacia seyal*
gurlo *Auchenoglanis biscutatus*
hoore free-born person
jaatigi host, especially of seasonal workers
jakka Muslim tithe
jebe *Lawsonia inermis*
jeeri area inland from the river
jinna evil spirit
kagume household head
kande basket loosely woven of thin branches, holding about twenty kilos of unthreshed grain
kenŋe a variety of coarse grass
keye rônier
kiide baobab
kolangal the River's floodbed
kome descendant of slaves
laxaasara late afternoon prayers
lawbe woodworker (Wolof)
mange warrior
melle fencing or roofing panel of coarse grass woven on a framework of branches
moodi Muslim cleric

mulunqu a variety of cucurbit
muno river spirit
muudu measure of grain, three to four kilos
ñecce coarse-ground sorghum or maize
ñobugu variety of sorghum
nigife one-quarter of a *muudu*
sakinbaya *Distichodus rostratus*
sakke woodworker
sallifana early afternoon prayers
sanbe *Grewia bicolor*
sanme variety of sorghum
sexenne *Balanites aegyptiaca*
sonbi porridge of coarse-ground sorghum or maize
suma variety of millet
surga seasonal worker
taalibe Koranic student
taarixu chronicle (Arabic)
tallaqe *Clarias gariepinus*
tefe *Combretum glutinosum*
tege blacksmith (man) or potter (woman)
tubab European, white person
tunbe *Garinarium excelsum*
Tunka ruler of Gajaaga
tunkan-yugo king
waate perforated stand used to leach potash from wood ash
worowolle *Striga*
xaame *Guiera senegalensis*
xaralenma younger Koranic pupil
xiile *Bauhinia rufescens*
xoole creek-bed
xooxa *Synodontis clarias* or related species
yugo man

Index

Aberdeen University 124, 152
Abidjan 85, 90, 242
Africa 7, 34, 96, 109–11, 124–5, 170, 259, 272, 278, 286
Agro-Action Allemande 281
Al-Fallah 241–2, 244, 256
Algiers 78–80
American 79, 120, 138, 140, 166, 243, 252
Amourousse, Monsieur 87, 104
Ancerville 100, 102
après-barrages 208, 226, 228, 236–7
see also dams
Aprin, Robert 112, 114–17, 120–30, 135, 143, 145, 154–5, 171
Arab 13, 15, 91, 109, 272
Arundu 10, 20, 24, 117, 120, 259–60
farming group 152, 157, 182, 183–7, 210–11, 236, 240, 243, 262, 282
Avignon 87

Ba, Thierno 236
Baalu 4–5, 20, 43, 68, 117–18, 181–2, 185–6, 251, 258
farming group 151, 157, 182–7, 194–6, 211
Bacili 4–6, 9–20, 23, 26, 28, 39, 41, 43, 45–7, 50, 69, 76, 82–3, 94, 99, 120, 137, 243, 252
Ibrahima Jaaman 82–3
Konko Goola 46–7, 50, 83, 99
as masters of the land 6–7, 10–11, 12, 20, 146
at war 14, 17–19, 23, 28, 43, 45, 47, 76
Bakel 6, 17–18, 23, 25, 39, 43, 47, 73, 82, 101–3
see also *commandant*; *préfet*; *préfecture*
as administrative centre 79, 82, 97, 101–2, 128, 133, 136, 145, 154, 190, 227
area 113, 155, 174, 185, 189, 220, 232, 257, 278, 285
as colonial post 13, 19–20, 24–32, 34–5, 62, 64, 68, 72–3, 77, 79, 82, 100, 147
département 185, 204
farming group 119, 137, 152, 154, 157, 188, 233, 172, 282
préfecture of 29, 72, 182, 185, 193, 204, 276
préfet of 72, 82, 97, 145, 161, 163–165, 167, 182, 189–190, 192–193, 203, 211, 217, 230, 242–4; see also Cissé, Oumar; Sy, Bocar
SAED *Ingénieur-délegué* in 177, 187, 192, 206, 216, 221, 225, 248–9, 251–2, 256, 283
Balandier, Georges 171
Bamba-Thialène 181
Bambara 14, 17, 31, 33, 82, 209
Bathily, Abdoulaye 7, 125
Bema 6, 26, 31, 33, 64, 70, 72, 79, 102, 146, 151, 209
Bizerte 76, 79–81
blacksmith 49, 56
see also *tego*
Bloch, Peter 213, 215, 232, 234
Bompard, Christian 205, 220, 278
Bordeaux 85, 87
Bouguenais 251
Brazil 85
Brazza 92
Brazzaville 90, 92, 95
Briend, Roger 276
Brochowski, Mademoiselle Gabrielle von 261, 269
BUD-Senegal 107
Bugnicourt, Jacques 184, 188
Bulebone 27–28, 36, 61
Bundu 10–11, 14, 17, 24, 26–7, 30, 61, 73, 137, 286

Caisse Centrale de Coopération Economique 177, 201, 257
Caisse Française de Développement 276–7
Cameroun 91, 109
Canada 90, 101

Index

Canadian 171, 178
Carbonare, Monsieur 185–6
Carmer, George 229
Casablanca 79, 82
cattle 5, 6, 15, 21–2, 34, 68, 85, 273
　see also oxen; ploughing
Cellule Après-Barrages 270, 276
Centrafrique 92
CFCF 281
Chad 92
champ collectif 118–19
　see also collective field
Chasin, Barbara 155
chef de canton 40–1, 82
Cherbourg 81
China 90
Christian Aid 248–9, 252, 260–2, 267, 269, 272, 277, 281, 284
CIDR 112–13, 122–3, 126, 129, 134, 136, 142, 144, 155, 162, 170
CIONGLA 228, 281
Cissé, Oumar 165
Cissokho:
　Cheikh 122, 153, 184, 238, 257, 278; see also Gawusu
　Mamadou 181–2
Claire, Madame 87, 104
CNCAS 201, 220, 238, 249
　see also Crédit Agricole
Coles, Julius E. 270, 283
collective field 139–40, 144, 154, 157, 162–3, 165, 172, 226, 235
Collin, Jean 188
Colonel Vaizey 89
colonialism 47, 72, 287
commandant, Bakel 30–2, 34, 77–9, 82
　see also *préfet*
communauté rurale 190, 211, 214, 251, 258, 274, 284–5
Compagnie Sucrière Sénégalaise 107, 280
Conakry 85, 90
Congo 69, 75, 91–93, 97, 116, 148
co-operative 178, 180, 185, 190, 208, 211–14, 231, 247, 251, 284
Cornell University 175, 196
Corsican 86, 121
COSAB 281
cotton 15, 34, 66–8, 70
cowpeas 37, 48, 71, 146–51, 289

Crédit Agricole 218, 274, 285
Croix-de-Feu 144

DAGAT 193, 217
Dakar 43–4, 55, 57, 73–9, 82–3, 85, 90–2, 95–6, 100, 103, 146–7, 150, 159, 166, 174
dams 113, 176, 188, 197, 201, 217, 220, 237–8, 270, 278
Daramaane 8, 15
David, Pierre 14
Davidson, Basil 287
De Gaulle 77
Delacoste, Colonel 82
Delcourt, André 14
Delta, *see* Senegal River
democracy 47, 126, 218, 224, 273–5, 287
Denbankaani 262, 266, 272
development 111–12, 115, 123, 126–7, 138, 141, 154, 168, 171, 189, 196–8, 259, 270–1, 273–5, 278, 280–1, 287
développement administratif 133, 141, 170, 196, 198
développement paysan 111, 127, 170, 196, 198, 281
Dia:
　Mamadou 73, 91, 95, 97, 172
　Oumar Khassimou 186–7
　Sada 218, 229, 246–7
　Yaya 187, 206
Diallo:
　Cheikh Oumar 205–6
　Doctor 166
Digokori 31
Diop, Abdoulaye 126–127
Diouf:
　Abdou 187, 217, 218, 248, 258, 278; see also Senegal, President of
　Pierre 177
Dogo 148, 150–1, 159
Domaine National, *see* National Domain
drought 102–4, 108, 122, 158
Dumont, René 112, 175–6, 209, 287
Dunkirk 85

Edinburgh, University of 214
Elsner, Richard 111–12, 123–4, 135, 144
emigration 44, 103, 155, 180, 257, 259, 268, 272, 275, 279

Index

Emile Bertin 79–80, 108
ENDA 178, 181, 184, 186, 188
ENEA 119, 125, 127, 143
Enfidaville 79–81
English 13
ethnicity 208–9, 215, 217
European Community 261, 269
Evin, Claude 257

Fabre et Freyssinet 85, 103
Faidherbe, Louis 25, 31
falo 20–1, 37, 39, 46, 48, 51, 70, 146–51, 263
Fan-lenme 7, 17, 20, 27, 137, 210–11, 241, 278
Fanqanne 8, 16, 21–2, 36–9, 51–2, 54–6, 64, 98, 100, 102, 117, 120, 132, 145–51
Federation (Fédération des Paysans Organisés du Département de Bakel) 133–285 *passim*
Feenstra, G. E. 240–1, 244–5, 261
Fendagesse 54, 58
Figaro, Le 259
fish 6, 12, 35, 48, 53, 55, 71, 101, 270, 273–4
floodlands 31, 39, 48, 51, 145–51, 158, 259, 263–6, 270–4, 279
Fondation canadienne 187
Ford Foundation 229
France 44, 46, 50, 55, 71, 75, 77–8, 82, 90–91, 96–7, 100, 102–3, 108–10, 116, 120, 128, 150, 159, 272, 275
Franke, Richard 155
French 13–14, 19–20, 22, 29, 32, 76–8, 84, 95, 154, 182, 257
Frey, Colonel Henri 25, 30
Fula 17, 19, 27–8, 32, 56, 72–3, 78, 82, 208, 210, 245, 247–248
  see also Halpulaar, Tukulor
fuñanŋe 37, 53, 66
Fuuta 14, 17, 20, 22–3, 28, 34, 45, 47, 66, 262, 286

Gaabu 18, 24, 64, 70, 118, 258
Gaajigo 19, 48, 95
Gabon 92
Gajaaga 6–8, 14, 17, 20, 22–3, 28, 34, 45, 47, 66, 69, 76, 88, 286

Gakura 7
Gallaade 6, 12, 17, 69, 118, 282
  farming group 152, 157, 159, 188, 233, 240, 243, 250, 262
Galliéni 31
Gambia 14, 21–2, 64, 66, 149
game, wild 55, 101
Gamera 15
Gamo 26
Gande 6–7, 46, 99, 118
  farming group 151, 157, 159, 188, 191, 202, 233, 243, 250, 262
Gangaji:
  Daado 36, 38, 41, 43–7, 55, 73, 77–9, 97–9
  Yugo-Xase 45–8, 50
garanke 58, 242–3
Garsingide 33, 52, 58, 147, 263
Gawusu 122, 124, 154, 166, 178, 181, 186–7, 258, 276–7
  see also Cissokho, Cheikh
Gece 58, 60, 64, 145, 150, 259
Gemu 32
Général Mangin 103
Georges Elek 76
German 78, 80–1, 109
Geyi 11, 21, 33, 38–9, 52, 98, 146
  Badara 74–5
  Madi Moodi 38–42
  Muulu 6–8, 12, 39
  Salun 55, 83, 101
Ghana 9, 259, 287
Gidimaxa 4, 20, 26, 27, 30, 45–7, 248
Golomi 13, 16–19, 27–8, 40, 47, 64, 77, 79, 100, 108, 119, 259–60
Gori 31–2, 39, 132
Gorvan, Lucien 124–5, 128, 143, 145, 153
Goundiam, Ousmane 173
grain 10, 32, 44, 47–8, 53, 55, 58, 60, 64–7, 69–70, 76, 99, 102, 145–51, 200, 263–6, 279
GRDR 111–12, 182, 203–4, 206, 257, 276
Greiling, Frank 269
groundnuts:
  as commercial crop 74, 95–6, 107, 155
  as women's crop 15, 34, 44, 53, 55, 66–71, 99, 133, 146–151, 165, 264–5

Groupement d'intérêt économique
  (GIE) 205–7, 211, 214, 251, 280, 284
Gucube 6, 16, 31
Guérini, Mémé 86
Guèye, Lamine 77–78
Guinea 90
guinea-worm 36
Guinguinéo 73
Gunjuru 8, 10, 12, 22
Guñan 6, 16, 68, 72, 118, 209, 258
Guñan Yaamadu 54–5, 58, 145, 150
Gurel Hayre 6

Hailu, Abiy 269
Halpulaar 134, 209–10
Harza Engineering 244–5, 247, 252–5
Havre, Le 85
Hayire 26, 45, 80, 262
Hoggar 101
Hortibak 276
Howard University 283

ICCO 238, 249–250, 267, 281
IMF 177, 201
Independence 47, 91, 96, 286
India 88, 254
indigo 65–70, 146
Iraqi 248
irrigation:
  abroad 34, 38, 87–9, 102
  Bakel area, in 122–281 *passim*
  Senegal River, on 103, 113–14, 156
Islam 12, 24, 64–5, 126, 240–1
Ivory Coast 116, 147–8, 150

Jaabalu 24, 56, 70, 146, 148, 150, 209, 259
Jaafunu 17, 64
Jaagili 16, 18, 29, 31, 35, 39, 45–7, 189
Jaaxali 8, 10–11
Jaaxo, *see* Tanjigoora
Jacquemot, Pierre 215
James, Wyndham 177, 202
Javelot 100
Jawara 6, 30, 118, 120, 228, 230, 232, 246–7, 285
  farming group 152, 157–8, 178, 188, 202, 230–3, 239, 250, 262, 282
Jawara II farming group 241, 256

Jeanne d'Arc 76, 79
jeeri 33, 40, 52–60, 64, 173, 196, 258
  see also rain-fed farming
Jogunturo 16
Joola 209–10
Jules Verne 76–80, 100

Kaarta 17
Kamara, Mahamme 42–3, 126, 132, 138, 140
Kammera 7, 16, 20
Kane:
  Papa 119, 125
  Racine 270, 276
Kaolack 33, 74–5, 81, 146
Kayes 30, 43, 121
  see also Xaayi
Keniyu 7, 10, 14
Kër Mejebel 74–5
Khoi Lê 173
Kidira 24, 125, 183–4, 202
kolangal 11, 39, 47–8, 50–51, 55, 59, 69, 130
Koliyaajo 28, 47
komo 37, 50, 52, 58, 59–63, 76
  see also slaves
Kounguel 34, 36
Kulikoro 43, 83
Kuŋani 8–52 *passim*, 77, 99, 115, 142, 145, 169, 239, 259, 262
  farming group 116–17, 132, 135, 140, 146–51, 157, 166–7, 172, 195, 202, 243, 250, 262, 272, 282
  farmlands 7–8, 21–3, 38–40, 51–2, 57–8, 145–51, 258–66, 279
  labour migration 73–6, 108, 145–51, 167
Kuwait 285

LaamDo JuulBe 25–6, 32
Laani 5, 8, 14, 18
Laborde, Robert 124, 139–41
Lagos 85, 92, 96
Land Tenure Center 213, 215, 232
leatherworkers 39
  see also garanke
Lebret, Father 278
Le Gendre, Bertrand 259
Léopoldville 92–4
Levtzion, Nehemia 9

Liberté 76
Libya 184
Lugere 54, 58, 145, 202, 259

Maasina 3
Madagascar 83, 89, 97, 109, 134
maize 36–8, 44, 48–9, 51, 53–7, 85, 87–8, 99–100, 102, 114, 119–20, 140, 146–51, 155–7, 165–6, 173, 183, 188, 191, 263–6, 279, 289
malaria 143–4, 274
Mali 43, 45, 61, 74, 91, 109, 113, 121, 149
Mammadu Lamiina (Daraame) 22–31, 35–6, 130, 147
Manayeli 6, 18–19, 23, 45, 47–8, 50, 95, 118
  farming group 152–3, 157–8, 178, 188, 202, 233, 250, 262, 272, 282
Manga 9, 18
Manga-Sakke 18, 99
mangu 6, 17–18, 45–7, 50, 286
Mansis, Firmin 221
Marseilles 84–7, 90, 100, 109, 134, 149–50
Matam 138, 140–1, 155, 157, 162, 188, 249, 285
Matforce 159, 220
Mauritania 245, 247–8
Maxaana 5, 14, 17–19
Mecca 8, 25
merchant marine 77, 83, 109, 148, 158
Messageries Maritimes 84–5, 104
Miller, David 175, 196
millet 68–9
Minister:
  of Agriculture 285–6
  of Emigrants 259
  of the Interior 166–7, 175, 192–3, 204, 215
  of the Protection of Nature 229, 238
  of Rural Development 121–2, 127, 162, 167, 190, 208, 232, 257, 276, 278
  of Women, Children and the Family 284
Ministry:
  of the Interior 175, 178, 181, 186, 193, 204, 210, 219
  of Planning 270–1
  of Rural Development 173, 193, 204, 214, 216, 226, 231, 238, 276, 278
  of Social Development 236–237
Mission Française de Coopération 202, 207, 257, 281
Monde, Le 258–9
Montcalm 76, 79
Monteil, Charles 31
moodi 9–12, 18–19, 32, 36–39, 58–60, 75, 91, 141, 149
Moors 4, 17, 19, 23, 34–5, 47, 247–8
Mouvement des Acteurs de la Vallée 286
Mudeeri 6, 47, 228, 285
  farming group (I) 152, 188, 241, 262, 282
  II, V, VI 241
  III 235, 241, 246
  VII 241–3, 246, 256

Naples 87
National Domain 235, 260, 280
navy 76–7
N'Diaye:
  Djibi 230, 232
  Latyr 186, 189
  Manthia 218, 230, 246–7, 285–6
  Mazide 185, 269
Ndioum 268, 272–4, 284, 286
N'Dongo, Sally 215
New Agricultural Policy 201, 204, 214, 218, 223, 229, 231, 252, 268, 271, 280, 285
New Caledonia 84, 88, 90, 104
NGO 123, 178, 183, 205, 228, 237, 269, 272, 281, 284
Niamey 92
Niang, Babacar 166
Nieuwe Weme, Gerard 235
Nigeria, Northern 229
Ñaŋaane 5
  Seyidu 137, 151–3, 181–7, 194–6

OFADEC 182–7, 194–6, 206, 213, 270
okra 70, 119, 157
OMVS 113, 115, 159–61, 171–2, 177, 180, 184, 188, 197, 201, 280
ONCAD 96, 177
ORSTOM 143
Ould Daddah, Moktar 248

oxen 41–2, 100, 138
  see also cattle; ploughing
Oxfam 123, 169, 177, 202, 220–1, 277
Oxfam America 202, 229, 239, 244, 248, 255, 266, 277, 281
Oxfam Belgique 221, 281

Papata 58, 91, 145, 150, 202, 259
Paris 87, 91, 104, 108–9, 149–50
Park, Mungo 13, 15
Patterson, Bill 243, 246
plague 44
ploughing 100, 119, 155, 181
  see also cattle, oxen
Pointe-Noire 85, 90, 92, 101
Port-Etienne 78
potatoes, sweet 37, 89
potter 44, 55
Programme Intégré de Podor 236, 270

racism 84, 109
RADI 269
railway 34, 43, 83
rain-fed farming 39, 48, 52, 133, 145–51, 154–5, 261, 263–6, 273–4, 279
  see also *jeeri*
Reboul, Claude 178
Réunion, La 134
rice:
  irrigated 42, 120–4, 132–92 *passim*, 220, 263–6
  swamp 66, 68, 146
river, *see* Senegal River
RND 175
Robin, J. 83

SAED 103, 107–13, 121–285 *passim*
  Comité paritaire set up by 205–10, 222, 224–5, 247
  Director of 122–3, 129, 153, 156, 164, 167, 173, 192, 222, 227, 249, 286
  Ingénieur délegué, see Bakel
Saint-Louis 13, 19, 23, 121, 126, 140, 154, 159, 161, 207–8, 230, 236
Sakiliba:
  Maamu 43, 83, 101, 133, 158
  Seega 6, 16, 73
sakko 4, 18, 20, 45–6, 286
Salem-Murdock, Muneera 270, 272

Sall, Tamsir 178, 184, 186
Samori 59–60
Saŋalu 236
Sanba-Gawlo 145, 259
Sanba-Salu 20–64 *passim*, 120, 130, 132, 138, 140–1, 145–7, 258, 263, 272, 289
Sanqare, Siliman 60, 132, 150–1
Saracolets 13, 31
Sarakhollais 5, 25, 38, 179
SATEC 95, 107, 113
schistosomiasis 143–4, 274
Scotland 13, 125
Seamen's Union 85
Seegankaani 7, 29, 35, 101–2, 114, 119, 137, 140, 146–7, 157, 258, 272
Seegu 25
Segala 68
Sélibaby 100, 247
Senegal 95–6, 103, 107–13, 197, 236
  Government of 103, 134, 143, 162, 171, 185, 188, 216, 223, 226, 247, 254, 278
  President of 153, 260, 284; *see also* Senghor, Léopold Sédar and Diouf, Abdou
Senegal River Basin Monitoring Activity (SRBMA) 270, 272
Senegal River Valley 122, 125, 145, 171, 257, 259, 261, 267, 272–3, 279–81
  Delta of 107, 113, 238, 249
  Lower 137
  Middle 22, 125, 249, 285
Senegalese 74, 84, 93, 95, 125–6
Seracolets 14
Serawoollies 14
SERDA 142–3
SERDI 143
Shear, David 168
Sisoxo:
  Bokar 133, 137, 166
  Famaxa 101, 150, 258
  Waali Silamaxa 39, 43, 83, 101
slavery 15, 59–63
slaves 7, 13, 15, 18, 21, 23, 35, 41–3, 47, 59–63, 69–70
  see also *komo*
So:
  Amma 20, 22, 27–8, 30, 35–8, 60, 97–8, 100

Baraka 15–16, 19, 27, 60, 83
Buleeli Asiya 16–17
Buleeli Denba 3–8, 21, 39
Haaruna 16–17
Juumu Haaruna 21, 27, 35–6, 42–3, 55, 82–3, 130
Mammadu 83, 114, 146, 159, 202
Mammadu Denba 20–3, 26–7, 60, 146
Siixu 36, 44, 54–5, 75–9, 82–3
Soboku 16, 20, 35, 202, 220
Social Science Research Council 145
SODEVA 96, 113
Soleil, Le 113, 187–8, 278
Somankidi 155
SONADER 156
Soninke 56, 64, 74, 75, 95, 99, 134, 160, 162, 170–171, 174, 184, 194, 204–5, 208–10, 215–16, 287
  see also Saracolet; Sarakollais; Seracolets; Serawoollies
sorghum 16, 37–8, 44–5, 48, 51, 53–8, 70–1, 88, 99, 102, 132–3, 146–51, 154–6, 158, 165, 188, 192, 261, 263–6
Soroma 64, 99
SOS-Sahel 251
South America 84–5, 90, 97
Soxolo 8, 10
Surcouf 79
Suxangide 52, 64, 68, 149
Sy, Bocar 250

taalibo 26, 37, 39, 52, 58, 63–6
Tahiti 84, 90
Tambacounda 27, 37, 44, 73, 148, 161, 174, 178, 182, 189, 194, 205, 228, 248, 258
  region, Governor of 161, 163–5, 181–2, 185, 188–9, 193, 217, 229, 275–6
Tanjigoora 8–10, 20–1, 24, 51, 58–9, 61–3, 77, 126, 128, 132, 184, 263–4
Al-Haji Haamidu 39, 66, 98–100, 148
Al-Haji MaJaaxon Ba 91, 94, 130, 263
Al-Haji Modibo 120, 141, 143, 147–8, 184
Al-Haji Mpali 132, 202, 263–4, 282
Amara 18–19, 21–2
Buna Fasunte 32–34, 52, 147
Bundunko 21–3, 53, 58, 91, 130

Dawuda Haawa 128, 132, 149
Fode Mammadu 35, 37, 41, 43–4, 58, 63–6, 141
Muusa Kaba 29, 39, 61, 71, 130, 132, 149
Sanba Haawa 21–2, 27–30, 34–9, 42, 44, 60–1, 91, 148
Seexu Jomo 22, 24–5, 31–3, 35–7, 39, 65
tego 6, 39, 45, 47, 49, 67
Thiam, Amadou 185, 219
Thierry, Benoît 276
Timmera:
  Haamidu Saada 251, 260
  Saada Sire 36, 40–2, 82, 252
Tiyaabu 6, 8, 12–14, 17–18, 20–1, 23–5, 31, 36, 45, 47, 49–50, 82, 99–100, 117, 119, 125
  farming group 152–3, 157–9, 188, 233, 242, 262, 272, 282
Toulon 79, 81, 108
Touré, Sékou 90
tractors 201–5, 213, 215, 234
tubab 19, 23, 29–34, 45, 59–60, 78, 84, 89, 109, 120, 141, 147, 197, 243
Tukulor 174, 208
  see also Fula; Halpulaar
Tunka 4–7, 18, 20, 41, 46–7, 50, 69
tunka-lenme 17, 38–9, 45

Uccelli 110–12, 124
Umaaru, Seexu (Tal) 20, 22, 24
Union Générale des Travailleurs Sénégalais en France (UGTSF) 125
United States:
  Ambassador 283
  Chargé d'Affaires 185
United States Agency for International Development (USAID) 122–79 *passim*, 195, 201, 213–14, 228–9, 232, 235, 261, 268, 272, 278, 284
  Bakel Small Irrigated Perimeters Project 130, 136, 144, 168–9, 179–80, 189, 213, 217
  Director 168, 229, 270, 283; see also Carmer; Coles; Shear
  Irrigation and Water Management Project 223–5, 232, 240–1, 244–5, 252–5, 282–3
USE 270

vegetable growing 117–19, 135, 154, 166, 263–6, 276–7
Vietnam 173
Ville:
　d'Amiens 101, 104
　de Massanga 80, 84, 89, 103, 134
　de Tamatave 89

Wageningen University 235, 240, 245
Wane, Mikhael 207–9, 211, 220, 242–3
War:
　Algerian 91
　First World 34, 62, 80
　Moroccan 68
　Second World 76–83
War on Want 123–6, 135–6, 142, 144, 155–6, 162, 169–70, 221, 239, 248, 281
West African Rice Development Association (WARDA) 235, 240, 244–5, 249, 253

Wisconsin, University of 213, 215, 232
Wolof 74, 210
women farmers 44, 52, 66–71, 116, 119, 145–51, 157, 172–3, 238, 243, 250, 254–65, 267, 269
World Bank 177, 201
Worms, Compagnie 83, 88, 134

Xaaso 14, 17
Xaayi 4, 10, 20, 23, 43, 92
　see also Kayes
Xanyaga 64
xaralenmu 63–4
Xasonke 6

Yafera 16–20, 40, 64, 82, 259–60
　farming group 178, 210, 238–9, 243, 250, 262
Yelingara 30–1, 118
　farming group 152, 157–9, 188, 191, 233